DEER FARMING IN NORTH AMERICA

DEER FARMING IN NORTH AMERICA

THE CONQUEST OF A NEW FRONTIER

JOSEF VON KERCKERINCK ZUR BORG

PHANTER PRESS
Rhinebeck New York

Published by Phanter Press
Rhinebeck, New York 12572

LIBRARY OF CONGRESS CATALOG CARD NUMBER 87-90712
ISBN 0-9619032-0-1

Printed in The United States of America

To my Wife and Partner

ILLIANA

Only with her support, encouragement, and understanding
was I able to make the dream of the Lucky Star Ranch come true,
a dream that began in August, 1978

ACKNOWLEDGEMENTS

I want to thank my parents-in-law, Udo and Irmel van Meeteren for their support over the years. They have been a tremendous help, for which I am very grateful.

There have been many friends and helpers during the first difficult years at Lucky Star Ranch, when we were not yet sure if deer farming could become a success. On the ranch, we had help from Sean and Mary Connors, Paul and Joan Kimmis, Rex Halladay and Jimmy Hoar, who has now become the manager and whose wife, Michele, keeps the office running smoothly.

I have had a lot of help and good ideas from my German friends, the Rüssmann family, Germany's foremost deer farmers.

I must mention my old friend Todd Bruno, who was involved with the graphic design of the taxonomy, and in other areas of this book.

Another good friend, Terence Boylan, from Rhinebeck, has helped me a great deal in putting this book together. What would I have done without him, his computer, and his ideas?

The sixteen colored pages with the drawings of the different deer by W. Weber have been kindly made available by Kindler Verlag GmbH, in München, West Germany.

Many people from Germany, New Zealand, England and the United States have given me their help, and I am thankful to each and every one of them.

My special thanks to my coauthors, Dr. Mary Giaquinto and her husband, Dr. Frank Giaquinto (Seaway Veterinary Service, Cape Vincent, NY, 13618) for their tremendous assistance in the writing of this book. They have both helped in ways for which it would be difficult to express enough thanks.

They immersed themselves in a field which was, until recently, entirely new to them. Through enormous effort, they became "experts" in fallow deer health and nutrition, and have been a valuable and irreplaceable source of help and know-how to us and to the whole industry.

In addition to their professional help and expertise, Mary Giaquinto has also edited this book, and did a great job of putting my script into proper English. After all, English is not my native tongue, and without Mary, the way I sometimes express myself would have given some people a lot to smile about. Think of it this way: How would you like to write a book in German?

NOTE: To avoid confusion, the word "doe" (female deer) will be capitalized throughout this book.

Contents

Whatever you can do,
Or dream you can, begin it.
Boldness has genius, power and magic in it! (Goethe)

1
Introduction

A LOOK TO THE FUTURE FOR FARMERS

In order to overcome the problems facing agriculture in the United States, we will require a new strategy. A strategy that combines the aims of profitable growth, cooperation between government and the private sector, and the willingness of farmers, processors, and government officials to take risks.

Success will require innovation, but most farmers have failed to innovate in the production of new crops. Many farmers are unwilling or unable to search for different products or traditional products at a different level of quality. Instead, they cry for help and focus on government subsidies.

The government, on the other hand, seems neither willing nor able to inspire farmers to find new kinds of agriculture, nor to innovate in the production of new products that would meet the differing demands of modern taste. Instead, progressive farmers feel wrapped up in and strangled by red tape. Any emerging new ideas suffocate. There are new lessons we must learn. And lesson number one is: Nothing will ever be as it has been before.

An overwhelming milk surplus forced the government to implement a "Whole Herd Buy-Out Program" (Dairy Termination Program) for dairy farmers. The result: approximately 1.5 million dairy cows were dispersed, creating an unstable beef market and causing problems in the beef industry. Beef prices dropped to an all-time low. Government "help" for the dairy farmer proved to be a burden for the beef farmer. So called "solutions" such as this make it clear that we must be able to help ourselves. Several states and producer groups have developed new ideas combined with effective marketing strategies and advertising campaigns. Their aim is to increase the visibility of their products and, therefore, increase their market share. Farmers, processors, and marketing hierarchies will continue to face advancing technology, changing references, and evolving consumer demands. They, as well as government officials, must be encouraged to innovate and to cooperate.

New products of high quality and new approaches are needed to help family-sized farms to survive. Besides the usual farm product list which includes dairy, beef, poultry, fruits, and vegetables, changing consumer tastes are encouraging new products of select quality: extra lean beef, fresh, legal venison, wine, aquaculture and more. While new products cannot be developed overnight, products are now being imported from foreign countries that could very easily be produced in the United States. A bold step in the right direction is necessary to overcome traditional insular attitudes of United States producers. One of the best opportunities lies in the production of fresh, superlative quality, legal venison, which currently sells for an outstanding price.

If you examine the facts, it is almost unbelievable that nobody in this country has ever before thought of deer farming in the way other countries have. Thousands of farmers in Europe and New Zealand are making a living from deer farming while people in this country still think of venison as an exotic meat that is normally consumed illegally.

In the face of changing consumer tastes and demands, farmers will need to adjust, and they will need help and guidance in making these adjustments. They will need marketing innovators who are ready to lead with new products to satisfy the consumer who seeks a quality product and is willing to pay a good price for it. Direct marketing of high quality products may be the best strategy.

When did you last see an agricultural product for which the demand exceeded the supply? Legal venison is such a product. As far as I know, it is one of very few.

That is why this book about deer farming and venison production has been written. I have assembled as much information as I could, drawn from my experience and from a wide variety of literature on the subject. The sources are all listed in the bibliography and referenced in individual chapters, and I am as pleased to give those who worked so hard for this information their proper credit, as they must be to know that it is contributing to a worthwhile effort.

I collected as much as I could and mixed it with what I have learned and experienced through the last nine years as a deer farmer in the United States. Though some of this may have to be adjusted due to differing conditions such as climate and crops, I feel confident that it is a sound basis upon which to begin an endeavor such as this.

Deer farming is a young industry, and though I know a lot more now than when I started, I still don't have all the answers. But if we are all willing to learn from our mistakes, we can help to make it become an important factor in America's agriculture.

Farmed venison has a great future in the United States because fresh, legal venison is not readily available, and hunted venison cannot be bought legally.

Deer farming, as I have mentioned before, is very popular in Europe and New Zealand. West Germany, for example, has approximately 1,500 deer farms, with the herds ranging from five to 500 animals. Most are family farm operations, with venison as a source of extra income and other farm products still the mainstay. Wild game is legally marketable in West Germany, and yet the country still imports more than 20,000 tons of venison each year. The average German consumes almost two pounds of venison per year, and this puts it second only to beef. The per capita consumption in the United States cannot even be measured in tenths of an ounce.

Whether or not deer farming will become a major factor in American agriculture will depend on the management systems that we can develop in the future. It may be that deer farming has the potential for a major contribution to world food supplies. At the moment, however, the main question is: how can we make such enterprises successful in order to help alleviate financial problems on the family farm? How may farm families survive and keep their farms with such supplemental income? After all, at least one-quarter of all family-sized farms in the United States are in financial trouble and it should be worth the effort to prevent further default. David W. Gross, extension specialist for economic development at Cornell University (Ithaca, N.Y.), invited a group of people to Cornell for a seminar about "Enterprise Development in Natural Resources" in September, 1985. I was honored to have been invited as a representative of United States deer farmers. I was shocked by what I learned about the income situation on the family farm. In December of 1986, I attended, as a speaker, the ADAPT 100 Conference in Des Moines, Iowa, initiated by *Successful Farming* Magazine, where nearly 6,000 farmers gathered to find new solutions to our threatening agricultural situation.

It is my hope that this book might advance the idea of deer farming throughout the United States and perhaps save some of those family farms threatened by bankruptcy. If I could help to prevent just one family from losing its farm to foreclosure, I would be happy.

I am confident, however, that the idea of deer farming combined with the right marketing approach can give hundreds of farmers the chance to stay on their land and make a decent living.

New Zealand has more than 2,000 deer farms, with mainly red deer and wapiti (North American elk), and also hybrids of the two. It is estimated that there are more than 300,000 deer in New Zealand, and according to statistics for the years 1985/86, New Zealand farmed approximately 18,000 fallow Does to fawn. The popularity of red deer and wapiti stems from the fact that New Zealand was mainly interested in harvesting velvet (the soft antler) and selling it to China and Korea. This market has waned, as oriental supply has better met its demand. Our main goal is the raising of deer for venison and, therefore, the fallow deer is an excellent choice. (Among gourmets, fallow and roe deer are both esteemed top quality. Being a concentrate selector, the roe deer is more difficult and expensive to keep than the fallow deer.)

There is vast experience among deer farmers in New Zealand. They have successfully introduced deer into modern farm systems and have taken a great step forward in the domestication of deer as a farm animal. I want to point out again that much of what I say in this book has been contributed by those farmers and experts whose knowledge is without a doubt greater than mine. Unfortunately, most of the literature from New Zealand concerns red deer and is, in most cases, not transferable to fallow deer farming. But many of the basic concepts are valuable and we should be grateful for everything we have learned from our New Zealand friends.

As founder and president of the North American Deer Farmers' Association, I am fully convinced that given today's state of farming expertise and the current market, fallow deer are probably the most suitable deer species to farm because they adapt extremely well to confinement and management. Fallow are efficient grazers and have a high conversion factor from grass to meat, much higher than most other farm animals. Many experts say that red deer have the better conversion factor, but conclusive proof is lacking. Another advantage of fallow deer is that young deer can be fed to appetite for maximum weight gain up to the age of 15 to 18 months without putting on fat, unlike sheep, which become fatty just over six months of age. Modern consumers don't want fat; why spend extra money to produce something that nobody wants? According to David Yerex, it takes twice as much feed to produce fat than to produce lean meat. Fallow deer also are readily available to farmers because they may be bought from almost any game park and they multiply easily, without major problems. Fallow are known to be resistant to disease and can, therefore, be kept in relatively small areas in close confinement with little danger.

Fallow deer have a relatively low purchase price for breeding stock. A good breeding Doe may sell for $300.00 to $450.00. That is a low price compared to the cost of red deer, whose hinds may cost up to $1500.00 or more.

For all these reasons, fallow deer have become increasingly important to deer farmers. But, most importantly, fallow deer venison is gourmet quality, especially if the animals have been farmed and are harvested in the right way at an age of 15 to 20 months. The controlled quality of the venison of those animals can hardly be equalled. It can only be approached by the quality of a red deer fawn that has been harvested at the age of six to eight months. The saddle (premier meat cut) of such an animal should not weigh more than eight to 12 pounds. If the saddle of a red deer weighs more than that, it comes from an older animal. In that case, the quality of the venison will not approximate that of a yearling fallow deer.

To summarize:

> Fallow deer can survive cold winters and hot summers.
> They are resistant to disease.
> They have relatively long life spans, easily reaching 15 to 20 years.
> They are known for high fertility and easy fawning.
> Their concentrate feed requirements are comparatively low.
> Their conversion factor is high.
> They are good-natured.
> Their meat production is good with high yield.
> They supply top-quality venison.

Even though some basics of deer farming have been established, it still is a young industry in which theories in some areas of husbandry, health and nutrition are yet to be proven. We must rely on logical thinking and a basic knowledge of domestic animal management. That is why future deer farmers would be well-advised to review as much information as possible, which, again, is easily said but not so easily done.

Our basic knowledge is growing because of people like David Yerex, Dr. Peter Fennessy, Dr. Geoff Asher, and Dr. Ken Drew, the Ministry of Agriculture and Fisheries (MAF) with Farm Advisory Officer Julie McCall, all from New Zealand; Prof. Günther Reinken, Dr. Erhard Ueckermann, Prof. Dr. Hermann Bogner, Josef Rüssmann, and Prof. Dr. R. R. Hofmann, all from West Germany; William Doyle from Arthur D. Little, Inc., Henderson, Friedlich, Graf, & Doyle, Inc from New York; Dr. John Fletcher and his wife Nicola from Scotland, Donald and Norma Chapman and G. Kenneth Whitehead, from England. And, of course, many others.

These people have taken the time and been enthusiastic enough to assemble some facts about a new, high-potential, agricultural venture, and about deer in general. We can all profit from their knowledge, much of which is printed in this book.

As our know-how is expanding and changing every day, we must be willing to acknowledge that some of what has been said here may have to be revised tomorrow. By the time this book is published, we might already have come to other views and better ways than the ones described here.

I like the words of Thomas Jefferson: "The most useful service we can render a culture is to add a new plant to its agriculture."

2

Phylogeny & Taxonomy

TAXONOMY

The taxonomy of deer on the following pages is not only of great importance to biologists but also important to deer farmers, in order to help them understand the relationship between the different species. In certain cases, it is an important tool in selecting the right species of deer to be farmed, and also to understand a little more about their habits and characteristics.

There are different opinions about how many subfamilies, genera and species exist in the *Cervidae* family. The number of subfamilies varies from five to seven, and genera from fifteen to seventeen. The number of species varies from thirty-one to forty, and the number of sub-species varies widely as well. From all the literature I have investigated, the delineation of the families and subfamilies of deer by G. Kenneth Whitehead makes the most sense to me. Of course, there will always be other versions, and there will be people with reasons for sticking to those versions. The deer family *Cervidae* consists of 7 Subfamilies, 17 genera and 40 species with more than 200 subspecies. Its main difference from other ruminants is the presence of antlers, not horns, on adult males.

The only exceptions are the musk deer and the Chinese water deer, the males of which do not grow antlers but have, instead, developed tusks as their primary offensive weapons. With the exception of reindeer and caribou, only males grow antlers. It happens, however, that female roe deer will occasionally grow small antlers, which do not affect their breeding capacity.

PHYLOGENY

KINGDOM	Animal	Plant
PHYLUM	*Chordata* (Vertebrates)	Invertebrates
CLASS	*Mammalia* (Mammals)	Amphibians Birds Bony Fishes Reptiles
SUBCLASS	*Eutheria* (Placental mammals)	*Prototheria* (Egg-laying mammals ex. anteaters) *Metatheria* (Pouched mammals/ marsupials, ex. kangaroos)
ORDER	*Artiodactyla* (Even-toed) *Ungulata* (Hooved)	*Perissodactyls* (Odd-toed ungulates) Horses Rhinos Tapirs Elephants Lagomorphs Rodents Armadillos Bats Carnivores Sea Cows Shrews Whales
SUBORDER	*Ruminantia* (Ruminants)	Nonruminants
FAMILY	*Cervidae* (Deer)	*Trangulids* (Mouse deer) Other true ruminants: *Antilocaprids* (Pronghorns) Bovids Giraffes
SUBFAMILY	*Cervinae* (Eurasian deer)	
GENUS	*Dama*	
SPECIES	*Dama dama*	
SUBSPECIES	*Dama dama dama* *Dama dama mesopotamica*	

There are seven Subfamilies:

Moschinae	Musk Deer
Muntiacinae	Muntjacs
Odocoileinae	American Deer and Roe Deer
Cervinae	Eurasian Deer
Hydropotinae	Water Deer
Alcinae	Moose
Rangiferinae	Reindeer (Caribou)

WHITE-TAILED DEER

Odocoileus virginianus

O.v. *virginianus*	Virginia USA, other States
O.v. *borealis*	South-east Canada, north-east USA
O.v. *dacotensis*	Alberta to North Dakota
O.v. *ochrourus*	North-west USA, Canada
O.v. *leucurus*	Oregon and west coast
O.v. *couesi*	Arizona
O.v. *texanus*	Texas, other States
O.v. *macrourus*	Kansas, other States
O.v. *taurinsulae*	Bulls Island
O.v. *venatorius*	Hunting Island
O.v. *hiltonensis*	Hilton Head Island
O.v. *nigribarbis*	Blackbeard Island
O.v. *seminolus*	Florida
O.v. *clavium*	Florida Keys
O.v. *osceola*	North-west Florida
O.v. *mcilhennyi*	Louisiana
O.v. *carminis*	North Mexico
O.v. *miquihuanensis*	Central Mexico
O.v. *sinaloae*	Mid-west Mexico
O.v. *veraecrucis*	East Mexico
O.v. *mexicanus*	Central Mexico
O.v. *acapulcensis*	South-east Mexico
O.v. *oaxacensis*	South Mexico
O.v. *toltecus*	South Mexico
O.v. *thomasi*	South-east Mexico
O.v. *yucatanensis*	Honduras
O.v. *truei*	Nicaragua, adjacent States
O.v. *nelsoni*	South Mexico, Guatemala
O.v. *chiriquensis*	Panama
O.v. *rothschildi*	Coiba Island
O.v. *tropicalis*	West Colombia
O.v. *goudotii*	Colombia, western Venezuela
O.v. *curassavicus*	Curacao Island
O.v. *margaritae*	Margarita Island
O.v. *gymnotis*	Venezuela, the Guianas
O.v. *cariacou*	French Guiana, northern Brazil
O.v. *ustus*	Ecuador (Andes)
O.v. *peruvianus*	Peru

MARSH DEER

Blastocerus dichotomus

Blastocerus dichotomus	Central Brazil to northern Argentina

MULE DEER / BLACK-TAILED DEER

Odocoileus hemionus

O.h. hemionus	West & central North America
O.h. sitkensis	British Columbia (+ islands off B.C.)
O.h. columbianus	B.C. and northern California
O.h. californicus	Mid California
O.h. inyoensis	California
O.h. fuliginatus	California
O.h. eremicus	North-west Mexico, Arizona
O.h. crooki	North Mexico
O.h. peninsulae	Baja California
O.h. sheldoni	Tiburon Island
O.h. cerrosensis	Cerros Island, Baja California

HUEMUL or GUEMAL

Hippocamelus antisensis

Hippocamelus antisensis	Northern Andes

Hippocamelus bisulcus

Hippocamelus bisulcus	Southern Andes

PAMPAS DEER

Ozotoceros bezoarticus

O.b. bezoarticus	Brazil
O.b. leucogaster	Central South America
O.b. celer	Argentina (Pampas)

PUDU

Pudu pudu

Pudu pudu	Chile, Argentina

Pudu mephistophiles

P.m. mephistophiles	Ecuador, Peru
P.m. wetmorei	Colombia

ROE DEER

Capreolus capreolus

C.c. capreolus (european)	Western Europe, Asia Minor
C.c. pygargus (siberian)	Siberia, northern Asia
C.c. bedfordi (chinese)	Manchuria, China, Korea

BROWN BROCKET

Mazama gouazoubira

M.g. gouazoubira	Paraguay, North Argentina
M.g. pandora	Yucatan, Mexico
M.g. permira	Isla San Jose
M.g. citus	Venezuela
M.g. nemorivaga	South-east Venezuela, Guiana,
M.g. mexianae	Mexiana Island
M.g. superciliaris	Brazil
M.g. murelia	South-west Colombia, Ecuador
M.g. tschudii	Peru
M.g. whitelyi	Peru

CERVIDAE

The Family Of Deer

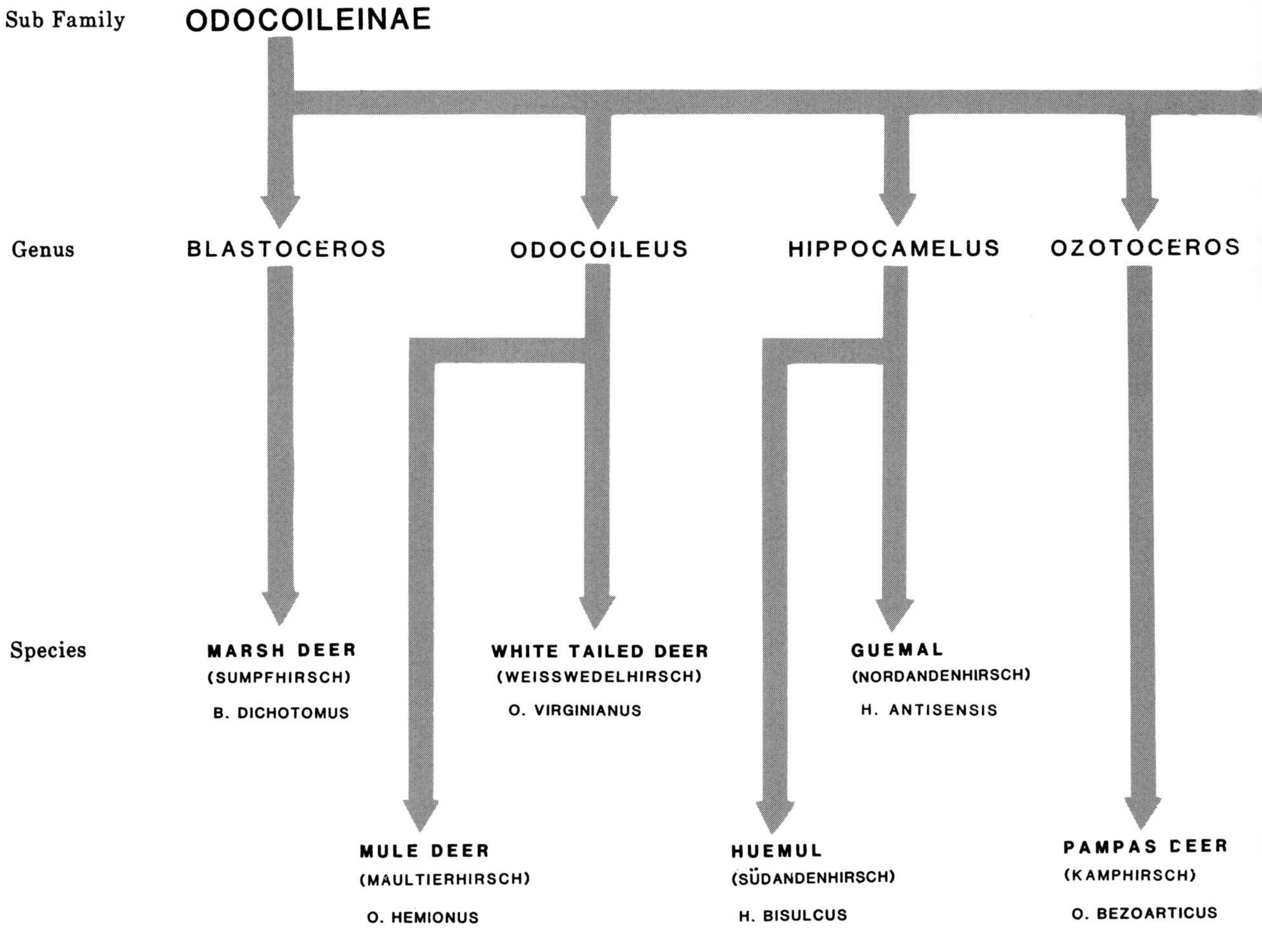

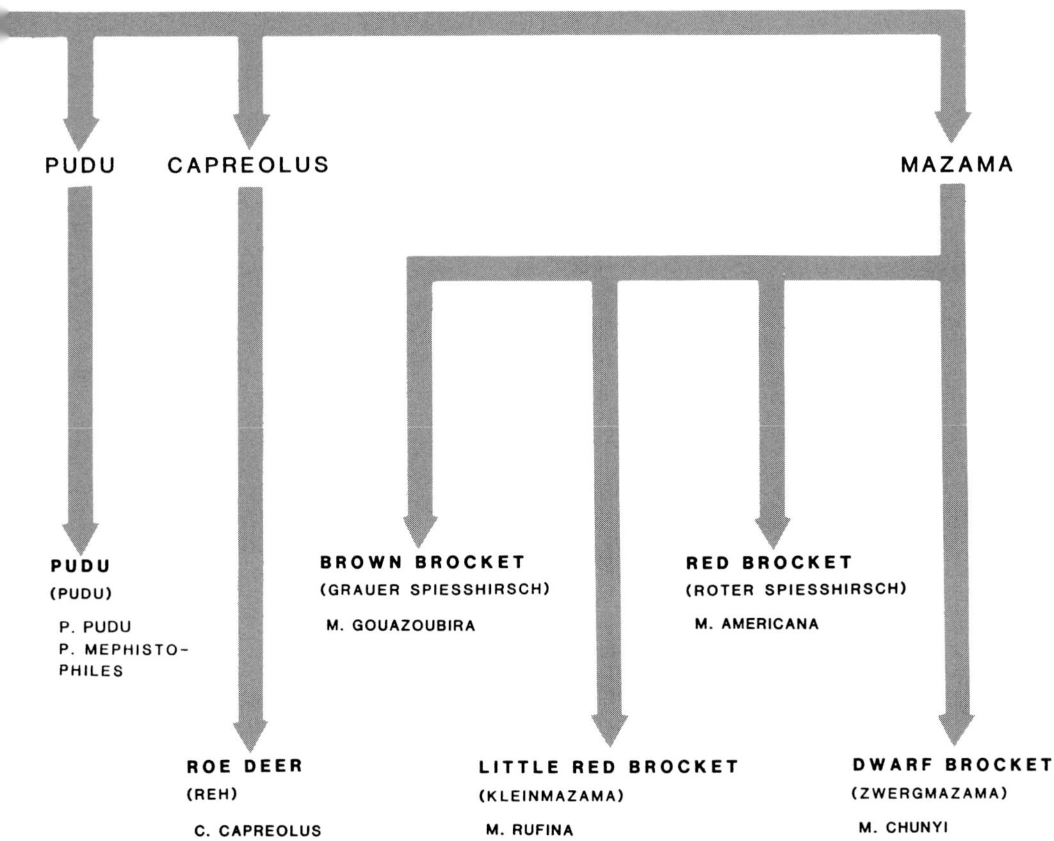

Kerckerinck/Bruno

LITTLE RED BROCKET

Mazama rufina

M.r. rufina	Ecuador, South-east Brazil, adjacent areas
M.r. bricenii	North Venezuela

RED BROCKET

Mazama americana

M.a. temama	Mexico
M.a. cerasina	Guatemala to Costa Rica
M.a. reperticia	Panama to South America
M.a. zetta	North Colombia
M.a. carrikeri	North Colombia
M.a. sheila	North Venezuela
M.a. trinitatis	Trinidad Island
M.a. americana	North-east South America
M.a. zamora	South-east Colombia to north-east Peru
M.a. gualea	West Ecuador
M.a. sarae	Bolivia, north-west Argentina
M.a. rufa	Paraguay, Argentina
M.a. rosii	North Argentina
M.a. jucunda	Brazil

DWARF BROCKET

Mazama chunyi

Mazama chunyi	Andes of north Bolivia, south Peru

MOOSE

Alces alces

A.a. alces	Northern Europe
A.a. cameloides	East Siberia, Mongolia, Manchuria
A.a. americana	Eastern Canada, north-east US of A
A.a. andersoni	Western Canada
A.a. gigas	Alaska, Yukon
A.a. shirasi	Wyoming, USA

WATER DEER

Hydropotes inermis

H.i. inermis	East China
H.i. argyropus	Korea

REINDEER / CARIBOU

Rangifer tarandus

R.t. tarandus	Norway to Russia
R.t. fennicus	European Russia
R.t. platyrhynchus	Spitsbergen
R.t. caribou	Canada, Alaska
R.t. groenlandicus	Greenland, Canada
R.t. granti	Alaska peninsula
R.t. pearyi	North-west Greenland
*R.t. dawsoni**	extinct
*R.t. eogroenlandicus**	extinct

* = extinct

TUFTED DEER

Elaphodus cephalophus

E.c. cephalophus	Southern China, northern Burma
E.c. ichangensis	Central southern China
E.c. michianus	South-east China

BARKING DEER

Muntiacus muntjak

M.m. muntjak	Java, South Sumatra
M.m. vaginalis	Northern India to south-west China
M.m. aureus	Peninsular India
M.m. malabaricus	Southern India, Ceylon
M.m. grandicornis	Burma
M.m. curvostylis	Thailand
M.m. annamensis	Indo-China
M.m. nigripes	Vietnam, Hainan Island
M.m. peninsulae	Malaya
M.m. robinsoni	Rhio-Linga archipelago
M.m. rubidus	North Borneo
M.m. pleiharicus	South Borneo
M.m. montanus	Sumatra
M.m. bancanus	Billiton, Banka Islands
M.m. nainggolani	Bali, Lumbok islands

FEA'S MUNTJAC

Muntiacus feae

Muntiacus feae	Tenasserim, Thailand

ROOSEVELT'S MUNTJAC

Muntiacus rooseveltorum

Muntiacus rooseveltorum	Indo-China

REEVES'S MUNTJAC

Muntiacus reevesi

M. r. reevesi	South-east China
M.r. micrurus	Formosa

BLACK MUNTJAC

Muntiacus crinifrons

Muntiacus crinifrons	East China

MUSK DEER

Moschus moschiferus

M.m. moschiferus	Northern India etc.
M.m. sifanicus	Western, central and southern China

Moschus sibiricus

M.s. sibiricus	Eastern Siberia, north Mongolia, Korea
M.s. sachalinensis	Sakhalin Island

Moschus berezovskii

Moschus berezovskii	Szechwan, China

NOTE: After deer family chart #2 on the following two pages, there are sixteen color plates showing most of the 40 species of deer.

All three deer family charts and vignettes of the color deer illustrations will be available as a 12 × 44 inch poster from Phanter Press.

CERVIDAE
The Family Of Deer

Sub Family	ALCINAE	HYDROPOTINAE	RANGIFERINAE
Genus	ALCES	HYDROPOTES	RANGIFER
Species	MOOSE (ELCH) A. ALCES	WATER DEER (WASSERREH) H. INERMIS	REINDEER/CARIBOU (RENTIER) R. TARANDUS

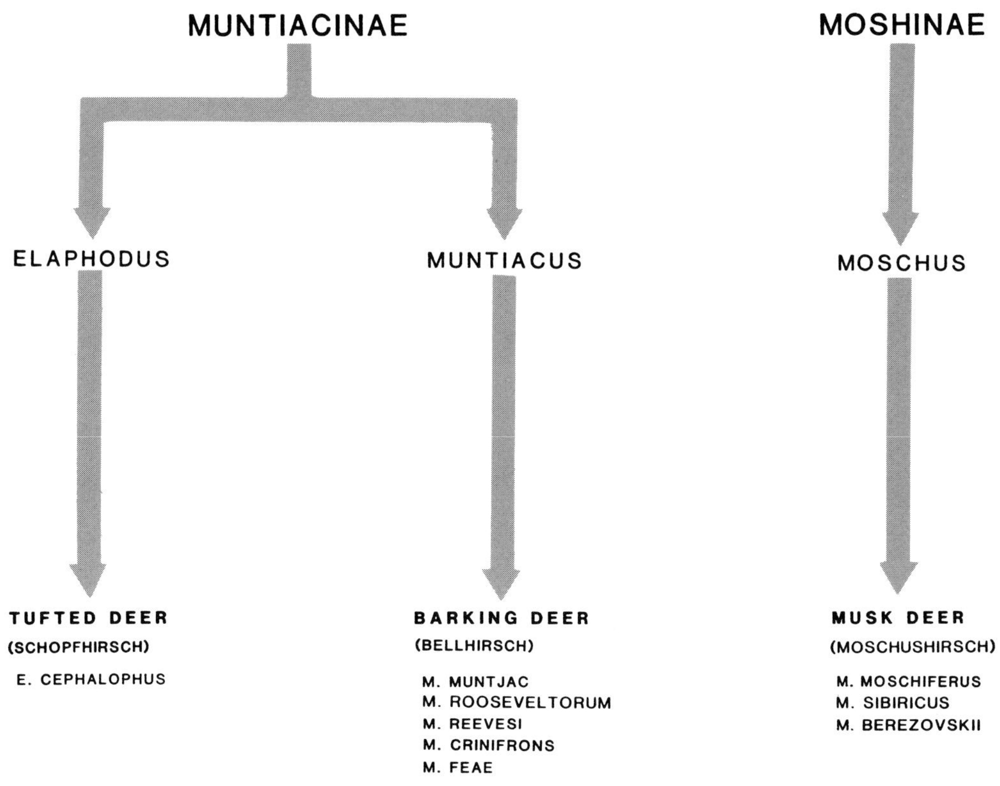

Kerckerinck/Bruno

PLATE I

Musk deer (subfamily *Moschinae*)
1. Musk deer (*Moschus moschiferus*)
1a. skull of a male with the long upper canines.

Muntjacs (subfamily *Muntiacinae*)
2. Javan muntjac (*Muntiacus muntjak muntjak*)
2a. skull of a male with long upper canines and small antlers on the high forehead.
3. Chinese muntjac (*Muntiacus muntjak reevesi*)
4. Tufted deer (*Elaphodus cephalophus*)

PLATE II

Fallow deer (subfamily *Cervinae*)
5. European fallow deer (*Dama dama dama*)
The colors vary from white, menil, red, brown, to black. Most variations are spotted.
6. Persian fallow deer (*Dama dama mesopotamica*)

PLATE III

Axis deer (subfamily *Cervinae*)
7. Axis deer (*Axis axis*)
8. Hog deer (*Axis porcinus*)

PLATE IV

Sambar deer (subfamily *Cervinae*)
9. Ceylon sambar (*Cervus unicolor unicolor*)
10. Sumatran sambar (*Cervus unicolor equinus*)
11. Rusa deer (*Cervus timorensis*)

PLATE V

Various deer (subfamily *Cervinae*)
12. Sambar deer (*Cervus unicolor mariannus*)
13. Alfred's deer (*Cervus unicolor alfredi*)
14. North Indian barasingha (*C. d. duvauceli*)
15. Schomburgk's deer (*Cervus schomburgki*)
16. Eld's deer (*Cervus eldi eldi*)

PLATE VI

Sika deer (subfamily *Cervinae*)
17. Formosa sika (*Cervus nippon taiouanus*)
18. Japanese sika (*Cervus nippon nippon*)
19. Dybowski's deer (*Cervus nippon hortulorum*), "Meihualu Deer" in China.
 a. male and female in their winter coat
 b. summer coated female with young c.

PLATE VII

Red deer (subfamily *Cervinae*)
20. Central European red deer (*Cervus elaphus hippelaphus*) male deer in winter coat, female in summer coat with young a.
21. Mediterranean red deer (*Cervus elaphus corsicanus*)
22. Barbary deer (*Cervus elaphus barbarus*)

PLATE VIII

Red deer (subfamily *Cervinae*)
23. Maral deer (*Cervus elaphus maral*)
24. Bactrian deer (*Cervus elaphus bactrianus*)
25. Hangual deer (*Cervus elaphus affinis*)
26. Altai Maral (*Cervus elaphus sibiricus*)

1a
♂
1
♀
2a
♂
2
♀
♀
3
♂
4 ♂

II

♀
♂
7
a
♀
a ♀
8
♂
♀

IV

V

17
♂
♀
18♂
19a♂
19a♀
19
19c

20 ♀
20 ♂
20 a
21 ♂
22 ♂

VIII

♀
27
♂
♂
28
a
♀

X

XI

b ♂
35 ♂
34
♀
♂
a

36
♀
♂
a
♀
37 ♂

XIV

43 ♂
♂
42
♀
♀
♂
39
♀
♀
44
♂

XVI

PLATE IX

Wapiti (North American elk, subfamily *Cervinae*)
27. Isubra Maral (*Cervus canadensis xanthophygus*)**
28. East wapiti (*Cervus canadensis canadensis*)
**"Malu Deer" in China

PLATE X

Various deer (subfamily *Cervinae*)
29. Thorold's deer (Cervus *albirostris*)
30. Pere David's deer or *Milu deer* (*Elaphurus davidianus*)

PLATE XI

Chinese Water deer (subfamily *Hydropotinae*)
31. Chinese water deer (*Hydropotes inermis*)
male and female with six young, 31 b. skull of a male with long upper canines.
Roe deer (subfamily *Odocoileinae*)
32. Sibirean roe deer (*Capreolus capreolus pygargus*)
33. European roe deer (*Capreolus capreolus capreolus*)
a. young, b. winter coat.

PLATE XII

White-tailed deer (subfamily *Odocoileinae*)
34. Virginia deer (*Odocoileus virginianus virginianus*)
a. young, b. in flight with erect tail
35. Key deer (*Odocoileus virginianus clavium*)

PLATE XIII

Black-tailed deer or Mule deer (*subfamily Odocoileinae*)
36. Black-tailed deer (*Odocoileus hemionus hemionus*)
37. Columbia mule deer (*Odocoileus hemionus columbianus*)

PLATE XIV

American deer (*subfamily Odocoileinae*)
38. Marsh deer (*Blastocerus dichotomus*)
39. See plate XV
40. Peruvian Guemal, northern Andes, (*Hippocamelus antisensis*)
41. Chilean Guemal, southern Andes, (*Hippocamelus bisulcus*)

PLATE XV

American deer (*subfamily Odocoileinae*)
39. Pampas deer (*Ozotoceros bezoarticus*)
42. Red brocket (*Mazama americana*)
43. Brown brocket (*Mazama gouazoubira*)
44. Southern pudu (*Pudu pudu*)

PLATE XVI

Moose (*subfamily Alcinae*)
45. Moose (*Alces alces*) a. young
Reindeer, Caribou (*subfamily Rangiferinae*)
46. North European reindeer (*Rangifer tarandus tarandus*)
a. wild, b. domestic
47. Caribou (*Rangifer tarandus caribou*)

FALLOW DEER
Dama dama

D.d. dama (european fallow deer)	Europe, Near East
D.d. mesopotamica (iranian fallow deer)	Iran

PÈRE DAVID'S DEER/ MI-LU
Elaphurus davidianus

Elaphurus davidianus	Imperial Hunting park, Peking

AXIS DEER or CHITAL
Axis axis

Axis axis axis	Peninsular India
A.a. ceylonensis	Ceylon

HOG DEER
Axis porcinus

Axis porcinus porcinus	North India, Burma, Ceylon
A. p. annamiticus	Thailand, Indo-China

CALAMIAN DEER
Axis calamianensis

Axis calamianensis	Calamian Islands

BAWEAN DEER
Axis kuhlii

Axis kuhlii	Bawean Island

SIKA DEER
Cervus nippon

C.n. nippon (Japanese Sika)	Kyushu Island, Japan
C.n. hortulorum (Dybowski's Sika)	Ussuri district, Manchuria
C.n. mantchuricus	Manchuria, Korea
C.n. mandarinus	Northern China
C.n. grassianus	Shansi district, China
C.n. kopschi	South-east China
C.n. taiouanus (Formosa Sika)	Formosa
C.n. pseudaxis	Vietnam
C.n. yesoensis	Hokkaido Island, Japan
C.n. centralis	Hondo (Honshu) Island, Japan
C.n. mageshimae	Mageshima Island, Japan
C.n. yakushimae	Yakushima Island, Japan
C.n. keramae	Ryukyu Islands, Japan

THOROLD'S DEER

Cervus albirostris

Cervus albirostris	Eastern Tibet, western China

SAMBAR DEER

Cervus unicolor

C.u. unicolor	Ceylon
C.u. niger	India
C.u. equinus	Burma, southern China, Sumatra
C.u. dejeani	South-west China
C.u. swinhoei	Formosa
C.u. brookei	Sarawak, Borneo
C.u. mariannus	Guam Island, Marianne Group (Probably extinct)
C.u. philippinus	Luzon, Philippines
C.u. alfredi	Central Philippines
C.u. barandanus	Mindoro Island, Philippines
C.u. francianus	Mindanao Island, Philippines
C.u. nigellus	West Mindanao Island, Philippines
C.u. apoensis	South-east Mindanao Island, Philippines
C.u. basilanensis	Basilan Island, Philippines
C.u. nigricans	Basilan Island, Philippines
C.u. boninensis	Bonin Island (probably extinct)

RUSA DEER

Cervus timorensis

C.t. timorensis	Timor, adjacent islands
C.t. moluccensis	Molucca islands
C.t. macassaricus	Celebes island
C.t. djonga	Muna and Buron islands (Celebes group)
C.t. floresiensis	Flores and lombok islands
C.t. russa	Java

RED DEER

Cervus elaphus

C.e. elaphus	Sweden
C.e. atlanticus	Norway
C.e. scoticus	Great Britain
C.e. hippelaphus	Europe
C.e. hispanicus	Spain
C.e. corsicanus	Corsica and Sardinia
C.e. barbarus	North Africa
C.e. maral	Asia Minor
C.e. hanglu	Kashmir
C.e. yarkandensis	Chinese Turkestan
C.e. wallichi	East Tibet
C.e. bactrianus	North Afghanistan, Russian Turkestan

WAPITI—NORTH AMERICAN ELK

Cervus canadensis

C.n. canadensis	Eastern North America (Extinct)
C.c. merriami	Arizona and north Mexico (Extinct)
C.c. roosevelti	Western North America
C.c. nelsoni	Western North America, except coast
C.c. manitobensis	Saskatchewan, Manitoba
C.c. nannodes (Tule Elk)	California
C.c. songaricus	Tien Shan Mountains
C.c. wachei	Western Mongolia
C.c. macneilli	Tibetan/Chinese border
C.c. kansuensis	Kansu, China,
C.c. alashanicus	South-east Mongolia
C.c. asiaticus	Altai to Transbaikalia
C.c. xanthophygus	Manchuria, Mongolia

BARASINGHA

Cervus duvauceli

C.d. branderi	Madhya Pradesh, central India
C.d. duvauceli	Northern India, south-west Nepal

ELD'S DEER / THAMIN

Cervus eldi

C.e. eldi	Manipur
C.e. thamin	Burma, Tenasserim
C.e. siamensis	Thailand, Indo-China, Hainan Island

SCHOMBURK'S DEER

Cervus schomburgki

Cervus schomburgki	Thailand (probably extinct)

CERVIDAE

The Family Of Deer

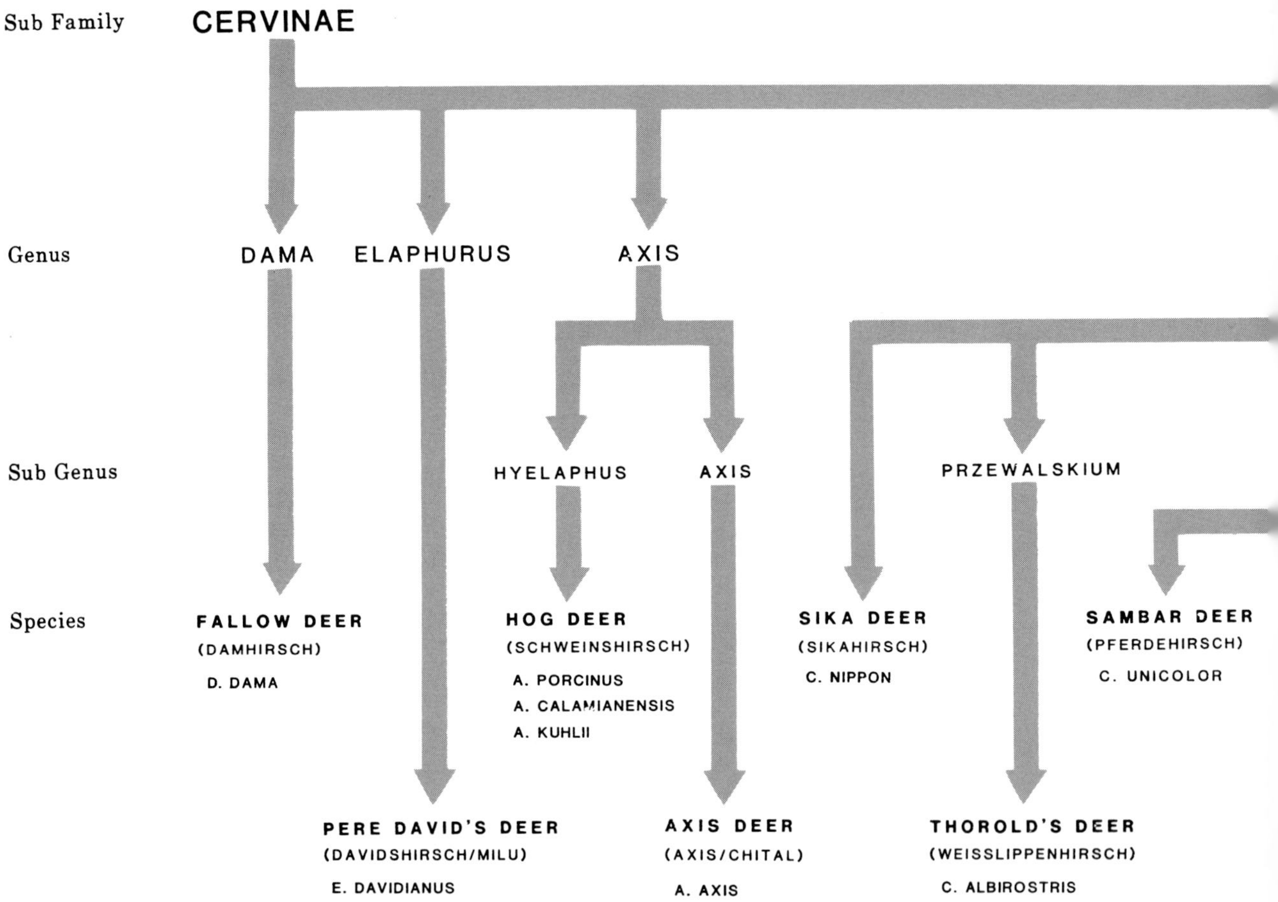

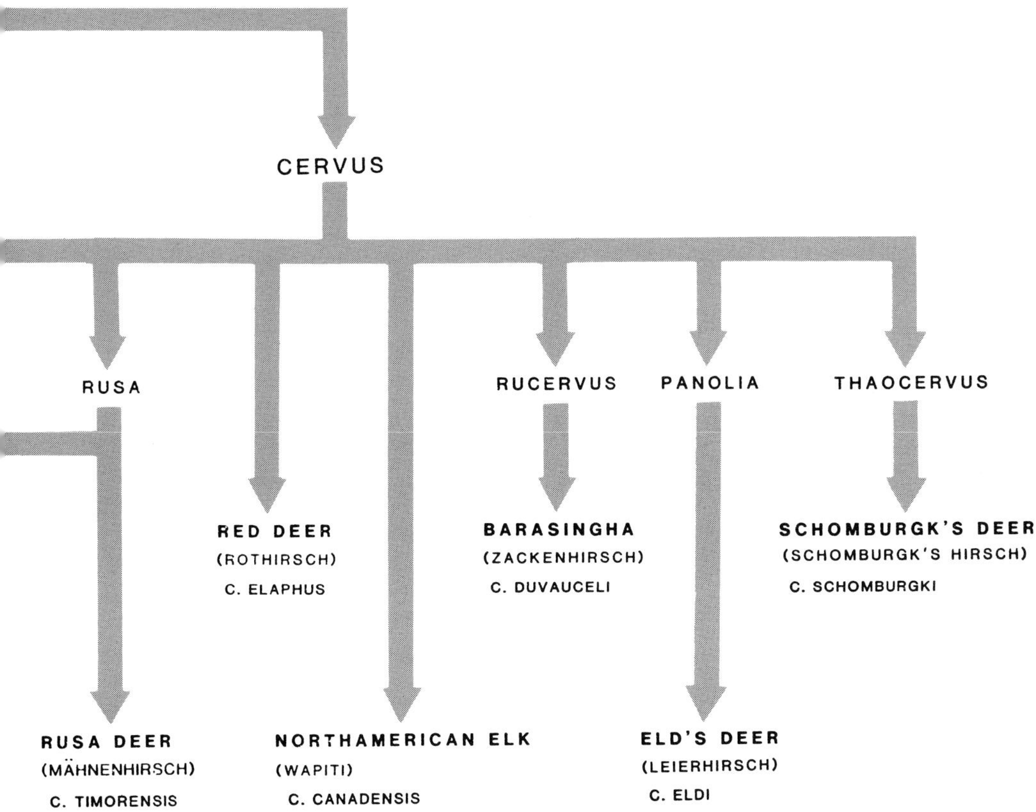

Kerckerinck/Bruno

3

History of Fallow Deer

Fallow deer form the genus *Dama* in the subfamily *Cervinae.* A main difference from deer of the genus *Cervus* is that mature *Dama* males have a broad, flat palm on each antler. They have no mane and adults usually have no upper canine teeth. Fallow deer are often confused with Japanese sika deer (*Cervus nippon*), because they are close in size and are often white-spotted. Male deer belonging to the genus *Cervus* have antlers with spike-like branches and possess a mane. They also have upper canine teeth in both sexes. (*Fallow Deer*, Donald and Norma Chapman).

The species *Dama dama* has two subspecies: the European fallow deer (*Dama dama dama*) and the Mesopotamian fallow deer (*Dama dama mesopotamica*), also called Persian fallow deer. Today, Persian fallow deer are nearly extinct, having a limited distribution in Southern Iran and Iraq.

Two years ago, I went to the Georg von Opel Game Park, in Kronberg/Taunus, West Germany, where there are about 30 Persian fallow deer left. This is probably the largest herd of Mesopotamian fallow deer in the world. They are higher in the shoulder than our fallow deer and approximately 40% heavier. They will cross breed with the European fallow deer, but the purchase price quoted to me during my visit was about $15,000.00 U.S. for a male fawn, plus transportation and quarantine expenses. Unfortunately, that was too expensive for my purpose.

Typical characteristics of fallow deer are the prominent Adam's apple, the long tail, and the white spots. They are sleek, long-legged, and appear quite stocky. The European fallow deer have a great variety of colors ranging from white to black. The Mesopotamian fallow deer show far less color variation. It seems certain that for millenia before Christ (perhaps as long as two million years), the European fallow deer lived all over Europe and the Mesopotamian fallow deer ranged in North Africa, Palestine, Syria, West Persia, and Mesopotamia. The evidence is insufficient to reach a definite conclusion as to whether fallow deer survived the last glaciation in Europe. Consequently, the origins of the European fallow deer will probably remain uncertain unless further fossil and archaeological specimens are found. (*Fallow Deer*, by Donald and Norma Chapman)

Most probably, the European fallow deer was reintroduced into the Mediterranean area by the Phoenicians between 1350 and 1000 B.C. The deer were used as sacrificial beasts for Baal. By that time, they were almost certainly semi-domesticated.

During Roman times, it was common for people to catch, tame, and harness deer; many writers left detailed accounts and records. Columella, in De Re Rustica 65 A.D. (*On Agriculture*), describes what amounts to our deer farming, so that "when the custom of giving feasts called for game, it might be produced as it were out of store." He goes on to describe in detail the best materials for enclosing the deer, giving them supplementary feed, etcetera. In Palestine, fallow deer were believed to be clean enough to be eaten.

Mandraki Harbor entrance, site of ancient colossus, Rhodes (*Dodecanese*) Greece

Photo: Courtesy of the Embassy of Greece

Mesopotamian fallow deer or Persian fallow deer (Dama dama mesopotamica) are nearly extinct. Some 30 still live at the George von Opel Game Park in Kronberg/Taunus, West-Germany, where this photo was taken.

Photo: E. Muller, Frankfurt

The presence of fallow deer at the entrance to Mandraki Harbor on the island of Rhodes, Greece, reminds us of the fact that fallow deer were brought there to fight serpents with their sharp hooves and to rid the island of poisonous snakes. Fallow deer were mentioned in the Bible, in "Deuteronomy", "(5) The hart, and the roebuck and the fallow deer and the wild goat, and the pygary and the wild ox and the chamois . . ." (Deuteronomy, 14:5) and also in the First Book of Kings . . . "Ten fat oxen, and twenty oxen out of the pastures, and an hundred sheep, beside harts, and roebucks, and fallow deer, and fatted fowl." (I. Kings, 4:23)

Jo Durden-Smith from *Connoisseur* magazine reported that fallow deer were found on a Mycenaean frieze (at Pylos, Tiryns) dating from the 13th century B.C., and that Apicius in the first century recommended cooking fallow venison with Jericho dates or dried plums. From the Mediterranean area, fallow deer were reintroduced into Northern Europe by the Roman legions who brought them to Spain, France, and, between 150 and 450 A.D., to England. The Romans also brought fallow deer to what is now known as Germany, from whence they spread all over the northern part of Europe.

The Kublai Khan, according to Marco Polo, had a form of game management for deer. Deer farms have existed in China since around 2,000 B.C. In Britain, poachers of deer were blinded or maimed, and the selling or buying of deer meat was banned. In 1671, the freedom to kill deer was further restricted to "qualified" people. Venison appeared only on the tables of the great and of the poachers. Because it was rare, it had a good reputation and was preferred to mutton. Venison's reputation as tough and strong emanates from the 19th century when it was common to hunt, and eat, mature trophy stags. Stags were often not killed on the first shot, and they were sometimes left outside, not dressed out, overnight. (Jo Durden Smith) Even William Shakespeare, as a young man, was supposedly caught poaching deer on the grounds of Chalcote Manor. (David Yerex) By Victorian times, deer had become "noble savages," and the habit of cooking them with wine, fruit and heavy, milky sauces arrived, as did the habit of marinating the meat for up to a week. The meat was old and tough because of the size of the deer shot. Venison, nonetheless, has remained a gourmet meat until today. (Jo Durden-Smith) In 1611, King James I had black fallow deer imported to England from Scandinavia. In Medieval times fallow deer were kept in many royal parks of Europe as a ready source of fresh venison. In both England and Denmark, fallow deer were kept in parks during the 12th century.

Captain John Smith, one of America's earliest settlers, admired "the great heapes of corn, venison, turkies and wild fowl" that the Indians had. The first thanksgiving with Chief Massasoit consisted of venison, duck, goose and (only probably) turkey. William Bartram, during the white man's first visit to the Creeks, was given "venison stewed with bear's oil, fresh corn cakes, milk and our drink, honey and water, very cool and agreeable."

In Schleswig-Holstein, Germany, fallow deer had been kept in parks long before they were released into the wild. The first historic import of fallow deer to Germany was from Denmark, when the King of Denmark, in 1577, sent 30 fallow deer to Landgraf Ludwig IV of Kurhessen, who kept them in a large park called Sababerg, near Kassel. The park was surrounded by a huge stone wall. (Ueckermann-Hansen)

The German name for fallow deer is "Damhirsch," which comes from "Denhirsch," meaning that those deer "Hirsch" came from Denmark. The English name fallow most likely comes from the word "fealo," the Saxon word for reddish-brown or dun. The Welsh name "gafr-damas" means Danish goat.

By virtue of their adaptability, fallow deer have spread widely throughout many countries. Although they do not find suitable living conditions in very high mountain regions, (max. altitude for Dama dama = 0–5000 feet) they exist (other than in the before-mentioned countries) in Sardinia, Italy, Spain, Portugal, Latvia, West Ukraine, Czechoslovakia, Hungary, Roumania, USA (Massachusetts–1932, Kentucky, Nebraska–1938, Texas), Canada, Argentina, Peru, Chile, North Africa, South Africa, Madagascar, Tasmania, Australia, New Zealand, Japan, Taiwan and Korea.

In West Germany today, there are more than 30,000 fallow deer in the wild. These deer, however, were released from parks; they were not originally wild and then taken into parks, as many people think. They were kept for the pleasure of the royalty and later became feral. They are not captured wild animals but are semi-domesticates (for a few thousand years), which have been released in great numbers, because they are desirable trophies for hunters.

4

Color and Figure Topographic Anatomy Weight Statistics

One of the most typical characteristics of fallow deer is the white spotting with which they are born. These spots do not disappear as they do on white-tailed deer or red deer, but remain throughout the deer's life.

For this reason, many visitors who come to the Lucky Star Ranch believe that all the deer they see are young animals. Considering the size of fallow deer compared to that of red deer, the inexperienced observer might still believe he is viewing an immature deer. With few exceptions, such as totally black or completely white deer, fallow deer remain spotted.

Unlike other deer, fallow deer show a wide color variation in both sexes. Some experts have counted as many as 14 different color variations. The colors range from WHITE (which is believed to be a mutation) to: BUTTERSCOTCH, LIGHT RED, DARK RED, REDDISH BROWN ("fealo"), BROWN, DARK BROWN, and BLACK (supposedly also a mutation). WHITE deer are normally butterscotch at birth and turn white at the age of eight to ten months. They are not at all albinistic; they have brown eyes. They were never appreciated much for hunting purposes but in a farm operation they do quite well.

We have, at times, suspected that the white deer might be a little more vulnerable to extremely cold climates and stress. They do appear taller and bigger than other deer, but that must be an optical illusion. Don't let them fool you. They are not heavier than other fallow deer.

Colored fallow deer carry their white spots throughout the winter. The rump and underside of the tail appear almost white. The light red, dark red, and reddish brown fallow deer have always been valued highly, because the deep red or russet red deer with prominent white spots, a dark stripe along the spine, and white dividing lines between the flanks and the belly, are the most attractive. The strong black stripe that shows along the spine, also called the dorsal stripe, is called the "Aalstrich" in Germany, "Aal" meaning eel and "strich" meaning stripe. In winter, the spots fade and the coat becomes a shade darker and a little duller. The rump patch is almost white. The brown and black fallow deer show this color all over their bodies except the legs and the belly, which are mostly grayish. Black deer sometimes have spots that are even darker than their coat but these spots are only visible at close quarters. In winter, their color turns close to grey. The rump patch is mostly grey. Black deer are neither smaller nor bigger than spotted deer. Some of them can be coal black.

All kinds of intermediate colors occur in fallow deer, but in all varieties the belly is of a lighter shade with a well-defined dividing line along each flank.

It appears that the velvet of dark-colored animals is darker and that of lighter animals is lighter in color.

The tail of the fallow deer is about eight inches long and constantly moving. It looks relatively long compared to that of other deer. It is mostly dark on top and almost white beneath and is surrounded by the white or gray rump patch, which in German is called "Spiegel," meaning "mirror."

If you breed black bucks to black Does, you will mostly get black fawns. The same will happen with white bucks and Does. The herd at Lucky Star Ranch is mixed, having all the different varieties. One year we used only spotted bucks and as a result 95% of our fawns were spotted. One breeder, who bought his first deer in 1979 through us from Jürgen Shulz in Catskill, NY, is now breeding his fallow deer in New York State according to color. He has three different herds which he named according to coat color: Black knights (black fallow deer); Queen deer (light red fallow deer, spotted); King deer (dark red fallow deer, spotted). If his method meets with success, he may find many breeders wanting special colors for special breeding stock.

From our experience on the Lucky Star Ranch, however, color has no influence on the size and health of the animals. We breed our deer for one reason only, and that is venison of the highest quality. Color does not influence that.

We are interested in tall, well-framed animals that have the ability to raise a healthy and strong fawn with a weaning weight that guarantees enough gain to reach minimum carcass weights of 60 to 70 pounds by the age of 15 to 20 months.

The figure of the female fallow deer appears to be light, smooth, and fragile. Most females are slightly higher at the hips than at the withers. This proved to be between one and two and a half inches in most cases. Height at the withers normally is between 30 and 40 inches; the length of the body, 51 to 68 inches.

The bucks appear heavy and stocky. The difference in hip height to height at the withers is not so obvious in the male as it is in the female. The buck's belly is strongly arched downward and the expressions "pot

In late summer, one can see that the bucks are developing rolls of fat around their necks. This is the time when the bucks are preparing for the rutting season.

Photo: Illiana Kerckerinck

bellied" and "paunchy" are appropriate. They stand about three feet tall and, at maturity, can weigh from 160 to 200 pounds.

In late summer, one can see that the bucks are developing rolls of fat around their necks. This is the time when the bucks are preparing for the rutting season.

They do not move much; they stay in groups, away from the Does. All they want to do is eat, rest, and gain weight and strength for the coming rut. During the rut, they will eat hardly anything and probably lose up to 40 pounds.

Strong hind legs and healthy hock joints are signs of a strong and healthy breeding buck who will do his job vigorously.

Prof. Dr. Günter Reinken points out an interesting detail about the fallow deer is a characteristic pattern on its nose. Directly above the upper lip between the nostrils, there is a pattern that will never change, equivalent in uniqueness to human fingerprints.

	Bucks 5 years and older	Bucks 1–4 years	Adult Does	Yearling Does	Fawns	
∅ Weights Prof Reinken W.-Germany	59.2– 74.2 kg. 130–163 lbs. ∅ *146.5*	38.6–49.2 kg. 85–108 lbs. ∅ 96.5	33.9–37.7 kg. 75–83 lbs. ∅ 79	26.8–31.4 kg. 59–69 lbs. ∅ 64	16.8–20.2 kg. 37–44 lbs. ∅ 40.5	
∅ Weights Lucky Star Ranch	135–185 lbs. ∅ *160.0*	102–108 lbs. ∅ 105	70–117 lbs. ∅ 93.5	53–88 lbs. ∅ 70.5	30–58 lbs. Females	37–67 lbs. ∅ 52 Males

∅ = Average

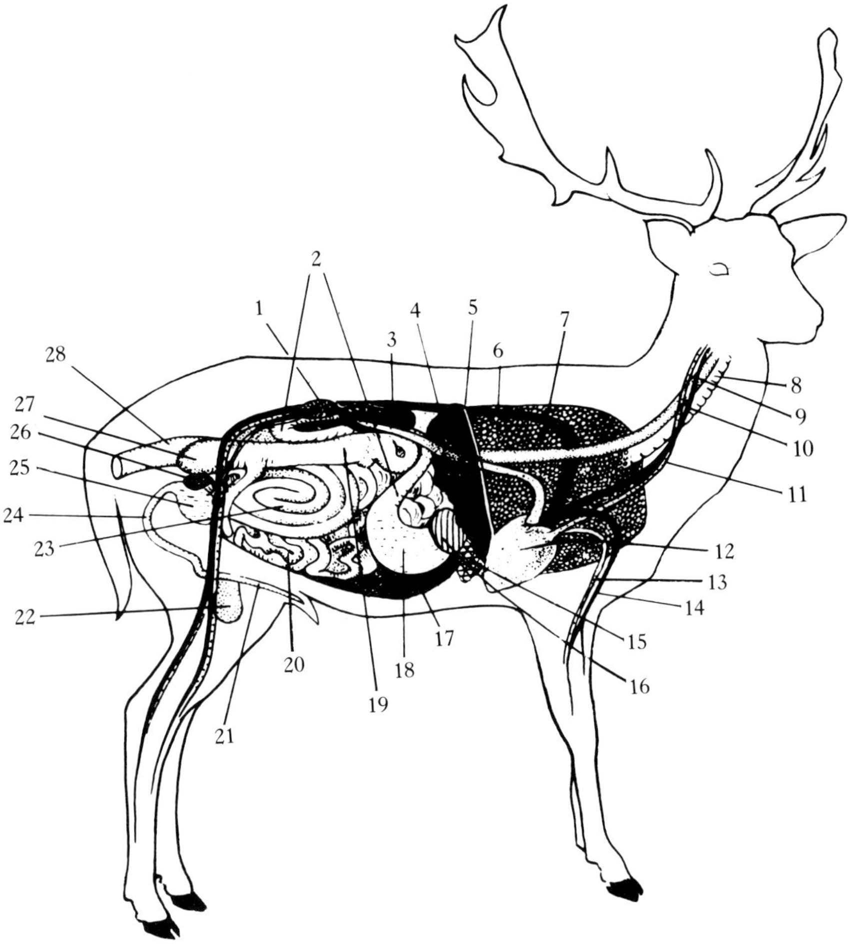

Topographic Anatomy. Courtesy of Dr. Erhard Ueckermann (Drawing by H. Osten)

1. Kidney
2. Duodenum
3. Spleen
4. Liver
5. Diaphragm
6. Aorta
7. Lung
8. Trachea (Windpipe)
9. Esophagus
10. Carotid artery
11. Jugular vein
12. Heart
13. Brachial vein
14. Brachial artery
15. Reticulum
16. Omasum
17. Rumen
18. Abomasum
19. Large Colon
20. Small intestine
21. Penis
22. Testicles
23. Spiral Colon
24. Urethra
25. Bladder
26. Seminal vesicle
27. Cecum
28. Rectum

5

Digestive System

(Reprinted from Fallow Deer, by Donald and Norma Chapman.

I am very grateful to Mrs. Norma Chapman and Terence Dalton Ltd., for their friendly permission.)

The outstanding feature of ruminants is the specialization of the digestive system. In the fallow deer, as in the other cud-chewing mammals, the stomach has four compartments. The rumen is the first and, in adults, by far the largest of these chambers. Abutting against the diaphragm and occupying most of the left side of the abdomen, it serves as a storage bag. The second and smallest compartment is the reticulum. The third chamber is the omasum which leads into the fourth, the elongated abomasum or true stomach, analogous to the single stomach of non-ruminants. The linings of the four stomach chambers are different, as connoisseurs of tripe will know. The rumen is densely lined with small finger-like villi and the reticulum has a network pattern of raised polygonal shapes: both these linings are tough enough to protect the underlying soft tissues from abrasion by the fibrous plant food. The omasum has folds resembling the pages of a book. All these devices increase the surface area over which absorption of nutrients can take place. The abomasum has a slight folding, giving way to an almost smooth lining at the duodenal end. Fallow deer consume large quantities of food which pass down the gullet into the capacious rumen. When the animal is at rest, this food is regurgitated, a bolus at a time. In the mouth the process of chewing the cud takes place. The vegetation is ground between the cheek teeth as the jaws move from side to side. When well-chewed, the bolus of food passes down the gullet but this time it is diverted to the omasum.

The functioning of a ruminant's specialised stomach depends upon the presence, principally in the rumen, of its microscopic flora and fauna. The vital role of these micro-organisms, which are bacteria and ciliated protozoa, lies in their ability to break down cellulose, the carbohydrate which forms the cell walls of plants, into glucose which is fermented in the rumen, producing enormous quantities of gases, principally methane and carbon dioxide. Obviously the gases cannot remain in the stomach: a complex and efficient mechanism of eructation ensures their removal. The gases pass up the gullet, into the back of the throat, some being forced down the windpipe into the lungs and absorbed into the blood. This re-cycling of some gas to the lungs and expulsion of the rest through the mouth is achieved almost silently, without any audible belch of appreciation for a good graze! If for any reason the gas is not released from the stomach chambers the animal suffers from bloat. A dead fallow deer soon becomes severely bloated, since fermentation and the production of gas continues in the rumen for a while after death but the eructation mechanism no longer operates. The gases therefore build up rapidly, causing great distension of the abdomen and making the skin taut, like a drum.

The density of the stomach's population of micro-organisms varies with the availability and type of food, and consequently with the time of year and with the physiological condition of the deer. A rutting red deer stag, which eats little, has only half the number of protozoa of a non-rutting stag, whereas lactating hinds have a far higher number of micro-organisms than a non-lactating female. Similar changes probably occur in fallow deer. Prins and his colleagues examined the rumina of 16 fallow deer shot in the Netherlands in the winter and calculated that the number of ciliate protozoa ranged from about 106 thousand to more than 2 million per millilitre of rumen contents. Over 80 per cent, and in some cases 100 per cent, of the protozoa were species of *Entodinium*. A few thousand million bacteria are also usually present in each millilitre of fallow rumen contents. These are digested and their proteins utilised by the deer. Thus a mighty fallow buck depends for his survival upon millions of single-celled organisms without which the vegetation he eats would be useless to him.

The relative sizes of the four stomach compartments vary with age in young deer. In a newly born fallow fawn receiving milk, the first three chambers are unimportant in digestion and therefore very small but the abomasum is comparatively large since it deals with the digestion of milk prior to its passage to the intestines. The changes in the proportions of the stomach chambers have been studied in a subspecies of white-tailed deer, similar in size to fallow. At birth, the rumen plus reticulum account for only ten per cent of the total stomach volume. Two weeks later this volume has increased fourfold and some microbial activity has begun. Fallow fawns of this age usually have some vegetation in the stomach as well as 'cheesy' milk. Between the ages of one and two months the rumen and reticulum of white-tailed deer increase tremendously, to account for 70 percent of the total volume. By this time the fawns are mainly dependent on rumen activity with milk forming only a small proportion of the total stomach contents. By four months the fawn's stomach chambers attain the same proportions as in the adult, when the rumen and reticulum together account for 85 per cent of the total volume. Although no precise measurements are available for fallow deer, the pattern of change and the relative proportions are similar to those that occur in white-tailed deer.

In the fifteenth century the liver, kidney and entrails, collectively known

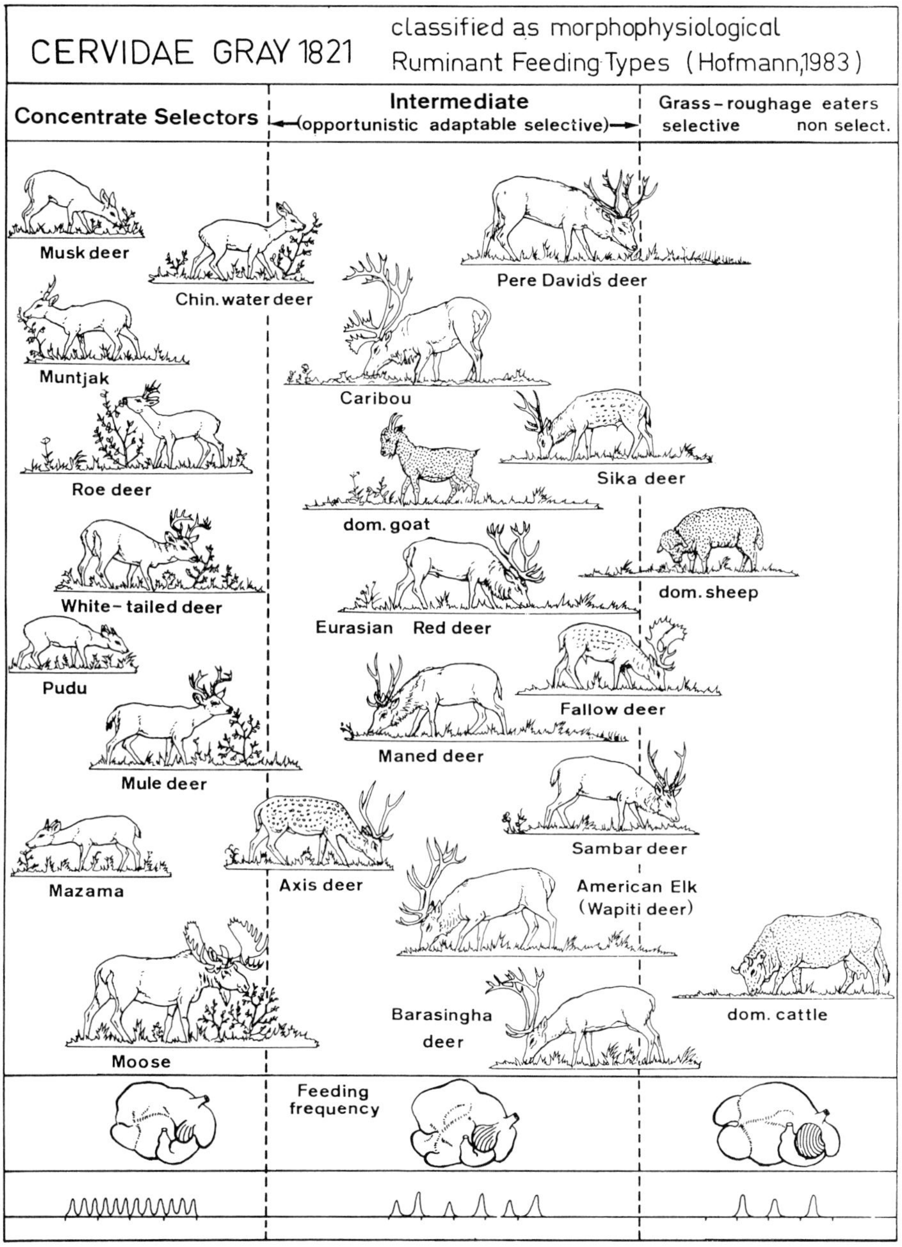

Courtesy of Professor Dr. R. R. Hofmann, University Giessen, West-Germany

as numbles, were made into a pie and served at the hunt feast to the lesser mortals whilst the nobility supped on haunch and saddle. The word numbles has evolved through umbles and humbles and survives today with the metaphorical eating of humble pie. The pancreas, testes and tongue were regarded as dainty morsels for the noblemen. The fallow has a large liver but there is no gall bladder, a feature shared by deer with a few other species of mammal including horses, rats and giraffes.

In the rectum the faeces are formed. These fewmets, as they are sometimes called, are cylindrical, firm, black, glossy and striated pellets which, when voided at the anus, may adhere to each other or form a string but usually fall in loose heaps or crotties of perhaps a hundred. Occasionally the faeces are soft and amorphous: Jackson found this type in the New Forest in May and June. On the ground, they frequently become covered with fungi and may look like a ball of fluff. Larsen studied fallow deer faeces in a park near Copenhagen and found that 55 species of fungi could occur on these droppings. The spores of at least 37 species were able to withstand passage through the gut and for some species this was essential before they could germinate.

In medieval times the condition of a deer was sometimes judged by its faeces, then called fumes, fewmishings or cotyings. It was customary for the huntsman to show the monarch a dish of fewmets, genteelly arranged amid a garnish of leaves, before he or she signified approval that the beast in question was worthy quarry, in prime condition.

This quaint ceremony may have been restricted to red deer stags for, in the fourteenth century, Gaston de Foix wrote about the fallow bucks "nor are his fumes put in judgement as those of the hart". Deer droppings still serve a useful purpose to the deer-watcher, for they are often the first signs of the presence of deer. The fewmets from an adult fallow Doe are about 15 mm long and 8 mm wide but the size varies; one end is pointed, the other rounded. Bucks have larger droppings which are also pointed at one end but often concave at the other. The droppings of many deer species are similar in shape and extreme caution must be exercised before trying to identify the species of a deer from its droppings. The fewmets of a fawn of one species may resemble, in size as well as shape, those of an adult of smaller species.

6

Scent Glands

(Reprinted from Fallow Deer, by Donald and Norma Chapman, with friendly permission from Mrs. Chapman and Terence Dalton, Ltd.)

SCENT GLANDS AND SENSES

The importance and significance of scent glands must be hard for man, with his relatively poor sense of smell, to appreciate although for generations he has exploited commercially the scents of some animals such as musk deer and civets.

The fallow deer has several scent glands, which are areas of specialised skin rich in secretory cells, and these were studied at the turn of the century by Tempel and by Zeitzschmann. These glands, the interdigital, the metatarsal and the suborbital, are all paired and occur in both sexes. In addition, bucks may have a scent gland in the penis sheath. Whilst the specific functions of these glands have not yet been demonstrated in fallow, they are probably a very important means of communication among members of the species.

The presence of interdigital or pedal glands has long been recognised: in medieval times the fallow buck and doe were described as beasts of sweet foot. At the base of each leg, in the mid-line immediately above the two cleaves of the hoof, is a fissure or narrow pocket in the skin. On the hind feet a pale yellow, soft waxy secretion, with a not unpleasant fatty-acid odour reminiscent of rancid butter, can be seen adhering to the hairs lining the pocket. The strength of the smell, as judged by the human nose, remains about the same throughout the year in both sexes. Our noses have not detected any smell from these glands in newly-born fawns but by the time a deer is two or three weeks old the glands are active. Interdigital glands are present also in antelope such as the sable, in which the gland opens to the outside by a central pore or duct. In fallow deer there is no central collecting duct, the secretion exuding from the skin. Microscopical examination shows that, below the thin epidermis, the upper part of the dermis has a regular pattern of areas of muscle interspersed with hair follicles with large sebaceous glands; the lower half of the dermis is composed of a mass of very large and much coiled sweat glands. The front foot also has a pocket but the skin is thinner and has less extensive glands, lacking the large zone of sweat glands present on the hind foot. The stain and odour associated with the hind feet are absent.

The position of the metatarsal scent gland is clearly marked externally by a raised oval cushion of paler hairs just below the hock, on the outer side of each hind leg. The cushion, about 40 mm long and 35 mm wide in an adult fallow, feels firm and has numerous very densely packed hairs which at their lower end are more or less embedded in a creamy coloured waxy secretion which has a faint fatty acid smell. A section through the gland shows that there is a deep zone of massive, sebaceous glands in the dermis, below which is a layer, of similar depth, of large, much coiled sweat glands. The glandular cushion is well defined in a newly-born fawn but appears to lack the scent of the adult.

Below the corner of each eye is a pocket in the skin which forms the scent gland known by a variety of terms which all indicate its position, namely, suborbital, preorbital, infraorbital or antorbital. Sebaceous and sweat glands are present in the skin lining the pocket, which is frequently filled with a brown wax.

Position of scent glands.

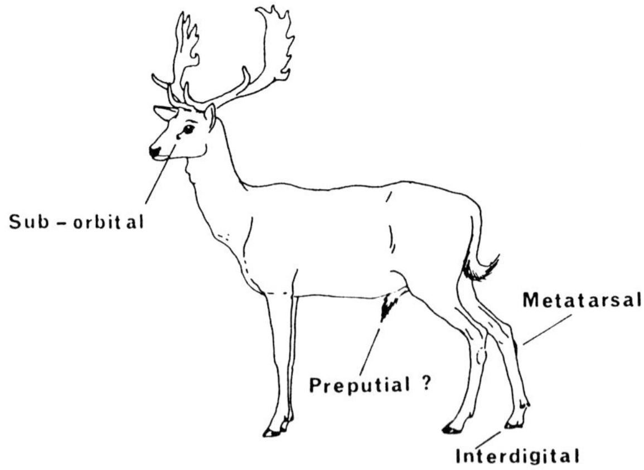

Courtesy of Dr. Diane Hughes

During the rut a buck has a very pungent and distinctive odour. A similar smell can be detected in the scrapes in the ground into which a rutting buck has urinated. Urine withdrawn from the bladder of a freshly dead rutting buck does not have this smell but the brush on the penis sheath reeks and becomes heavily stained during the rut. It seems that the characteristic smell is imparted to the urine either as it passes down the urethra or on contact with the air. The surface of the last 15 to 20 mm of the penis sheath is usually cream in colour and has numerous short, finger-like dermal papillae on the inner wall. Just before the rut the tip of the sheath becomes everted, so the papillae are on the outside and they become enlarged and black: the whole area becomes flattened and looks like cracked earth. Examination of an organ in this state, which looks as though it may be decomposing, must have prompted a comment which was the subject of a letter by Wherry in *The British Medical Journal* for September 1902, which said: Sir Thomas Browne corrects the vulgar error that the stag also sheds his 'pizzle' after the yearly rut . . . . it does not annually rot away. Fallow bucks are often called "stags", even now. The tip of the sheath reverts to its normal shape and colour soon after the rut. Adjacent to the blackened area, the skin of the penis sheath has large multilobular sebaceous glands and large sweat glands and the area appears similar to other skin regions specialised for scent production. It is thought that the substance responsible for the rutty odour of a billy goat is produced by the sebaceous glands and their secretion causes discolouration of the skin. It is known that in at least some mammals, androgens, the male hormones, cause an increase in size and activity of sebaceous glands. Since the testes of fallow deer are at their maximum activity at the time of the rut, the androgen level is almost certainly at or near its highest. The large size of the sebaceous glands in the penis sheath suggest that they may be associated with the production of the buck's rutty odour.

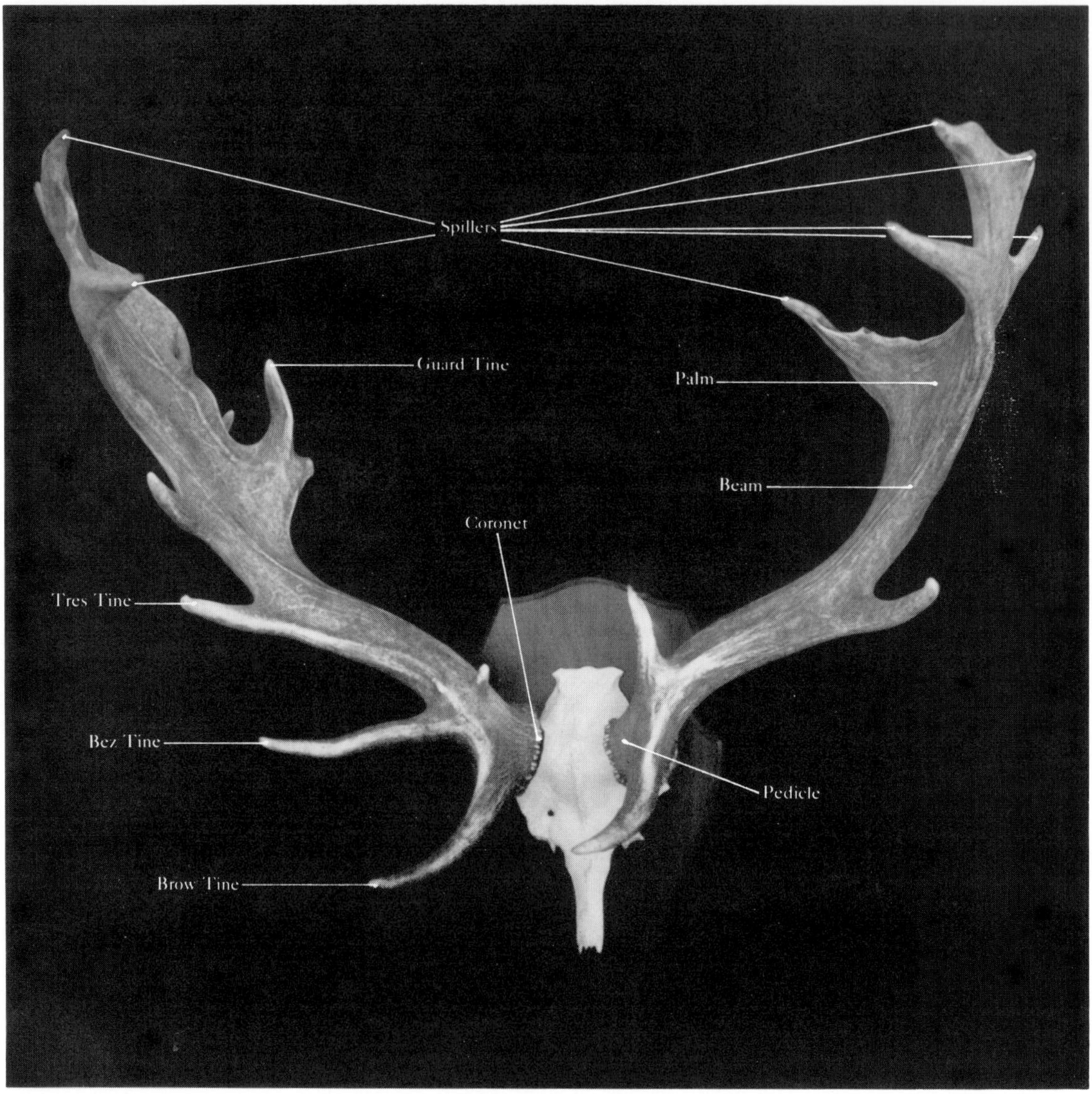

The antlers of a Lucky Star Ranch Fallow Buck with Brow Tine, Bez Tine and Trez Tine. The Bez Tines are often missing.

Photo: Lila Garnett

7

Antlers, Hair, Vision

One of the most impressive things about fallow deer are their antlers, which are proportioned completely differently from those of red deer or wapiti (North American elk).

Fallow antlers begin quite similarly with the "coronet" and have "brow tines," "bez tines," and "trez tines" and then shape into a unique palmated antler with many "palm points" (spillers) and "guard tines" at their lower end. The antlers of a fallow deer remind me of those of a moose. The "bez tines" are mostly missing.

Fallow buck in velvet, antlers not yet fully developed. *Photo: Author*

Three young bucks showing first signs of new antler growth

Photo: Author

Antlers are bony outgrowths of the skull. They are found in most members of the deer family and are normally replaced annually. With the exception of reindeer and caribou, only males grow antlers. Antlers are formed of living tissue covered by a finely haired skin—the velvet. The growing antler is well-supplied with blood vessels and nerves via the velvet. Initially, the antler is cartilage, which becomes mineralized and finally is converted to bone. At about this stage, the blood supply dries up and the velvet dies and is rubbed off.

Horn, on the other hand, consists mainly of keratin filaments, surrounding a bony core. (Keratin is a class of proteins that includes wool, hair, and hoof.) Horns are retained throughout life, except in the pronghorn (*Antilocapra americana*) of western North America, which casts its horn covering annually.

During the time of growth, antlers are covered with the velvet. There are many blood vessels and much nerve supply in this soft antler, which takes approximately 70 to 120 days to develop.

By the age of four months, the skull of a male fawn has developed a swelling—the pedicle, a permanent projection of the frontal bone. On the pedicle grows the coronet; old antlers shed from below this coronet. New growth arises from the skin around the pedicle.

Older bucks cast off their antlers at the end of April or the beginning of May. Development of the new antlers begins in May and takes until the end of August. At the beginning of September, bucks start to rub off the velvet. The antler is now dry and hard and becomes a weapon. The time between rubbing off the velvet and shedding the antler is about eight months

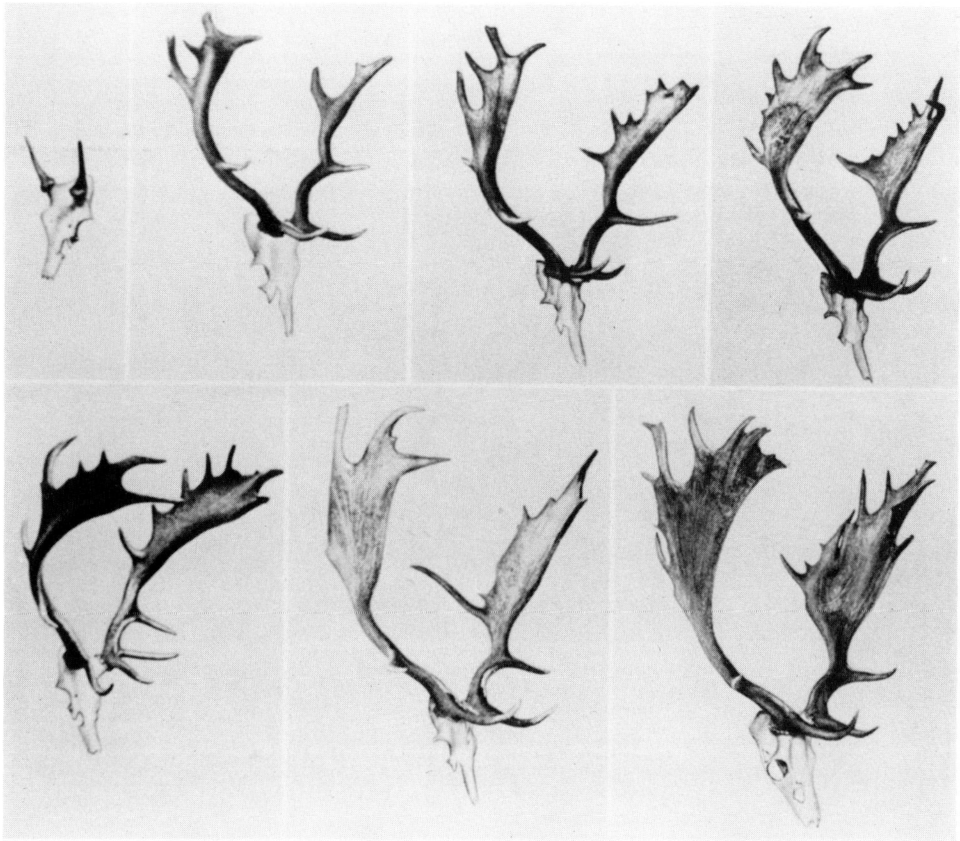

Antler development of fallow deer.
top row from left to right: 1. spike horn (first head) 2. second head, 16.5 inches long. 3. third head, 20 inches long, 2.8 pounds. 4. fourth head, 19 inches long, 3.3 pounds.
bottom from left to right: 5. head, 20,6 inches long, 6. and 7. head, 24.7 inches long, 4.8 pounds. 8. eighth head, 26.8 inches long, 6 pounds.

Photo: D. Lülfing, courtesy of Dr. Erhard Ueckermann

(longer than in most other deer species). There is an anatomical phenomenon of the antler in the area of the "trez tine," where the antler has a functional perforation, which serves as a preformed breakpoint. It is meant to prevent serious injuries of the skull by snapping at a maximum force. In other words, that part acts like a machine's sheer pin which will break before serious damage occurs and irreplaceable parts are damaged.

The fallow deer produces more annual mass of antlers in relation to its body weight and size than any other deer. The extinct giant fallow deer (*Megacerus giganteus*), with the misleading name "Irish elk" (elks have never been found in Ireland), had antlers which weighed up to 100 lbs., with a

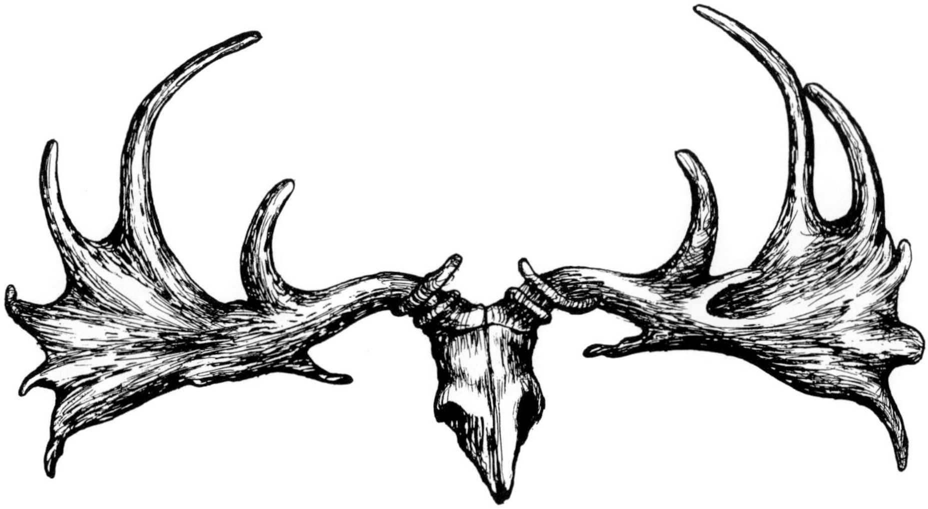

The antlers of the giant fallow deer (*Megacerus giganteus*)

(Drawing: Chuck Constantino)

span of eight to eleven feet. Caribou males and moose males are the two animals today with the largest antlers of all deer. A moose can carry antlers weighing up to 50 lbs.

The antlers of a mature fallow deer buck are desirable trophies to hunters all over the world, especially because these deer are extremely wary and are very hard to hunt. Unfortunately for our research, the antlers are not recorded in North American record books.

The antlers consist of:

10% Water
50% Ash(inorganic substance)
40% Organic Matter

The inorganic substance consists mainly of dicalciumphosphate with some calcium carbonate. The organic matter is mainly a collagen bound protein imbedded with minerals and cellulose. The antler of a yearling spike horn consists of:

10% Water
55% Ash
35% Organic Matter

(Ueckermann/Hansen)

VISION

Fallow deer have a kind of binocular vision, in contrast to the white-tailed deer, which seems not to recognize an immobile person from a distance greater than 50 yards (if the person holds still and the deer does not catch his scent). Because of the location of the eyes, more to the front of the head, fallow deer have overlapping lines of sight allowing them to see three-dimensionally. The fallow deer has a wide field of vision with a greater depth perception than most deer species. Fallow deer can see from straight ahead to far behind. A fallow deer can sight and recognize a person at distances of approximately 200 yards—and sometimes farther. In open fields, they have been observed to avoid a person walking towards them from 700 yards. In the dusk and at night, their ability to see is like that of average humans.

The olfactory sense (sense of smell) of fallow deer is not so well-developed as that of other deer. Fallow only seem able to smell a human at a distance of 200 yards or less (red deer may detect a human at 1,200 yards). Fallow must usually depend upon vision more than scent.

HAIR

Every spring and fall the fallow deer sheds its coat of hair in a molting process. The older an animal is, the later in the spring this change takes place. Young animals begin changing their hair approximately in early May in the northern United States latitudes. Older animals begin three weeks later.

In the spring molt, the regal appearance becomes shabby. The shedding takes about 40 days. During that time, the animals look bedraggled and patchy-coated. Accounting for variations in age, the summer coat is usually fully developed by the end of June.

If animals are in a state of poor health, are parasitized, or suffer from malnutrition or other deficiencies, such as lack of selenium or vitamins, the shedding process may stretch into July. The change of coat begins with legs, head and neck, then slowly the rest of the body changes its hair.

The quality of the summer coat and the length of time for the molt are good indicators of the health of an animal and the quality of the feeding program.

Because most of the Does are probably pregnant, shedding for them takes a little longer in spring than it does for yearling females, who are not pregnant yet. Molting also takes a little longer for the pregnant Does than it does for the bucks. In the fall, it is the other way around. Because the change of hair takes place during the rut, the bucks need longer because the rut is stressful. The bucks shed from early September to late October or even early November, which is approximately 50 days. The Does have already had their fawns and are not pregnant during that time, so they shed a little faster.

Deer have three different types of hair:

1. Tactile hair (feelers or whisker type hair)
2. Guard hair (bristle or spine type hair from primary follicles)
3. Wool hair (secondary hair, fine soft under hairs from accessory or secondary follicles—wool, fur or lanugo type hair)

In the summer coat, all three kinds of hair are shorter than in winter. The winter coat has more coarse hair and more secondary hair. The guard hairs are the longest and the secondary hairs become quite thick. Hanging just below the vulva, fallow Does have hairs forming a tuft, equivalent to the brush on the male's penile sheath.

Antler Growth Calendar

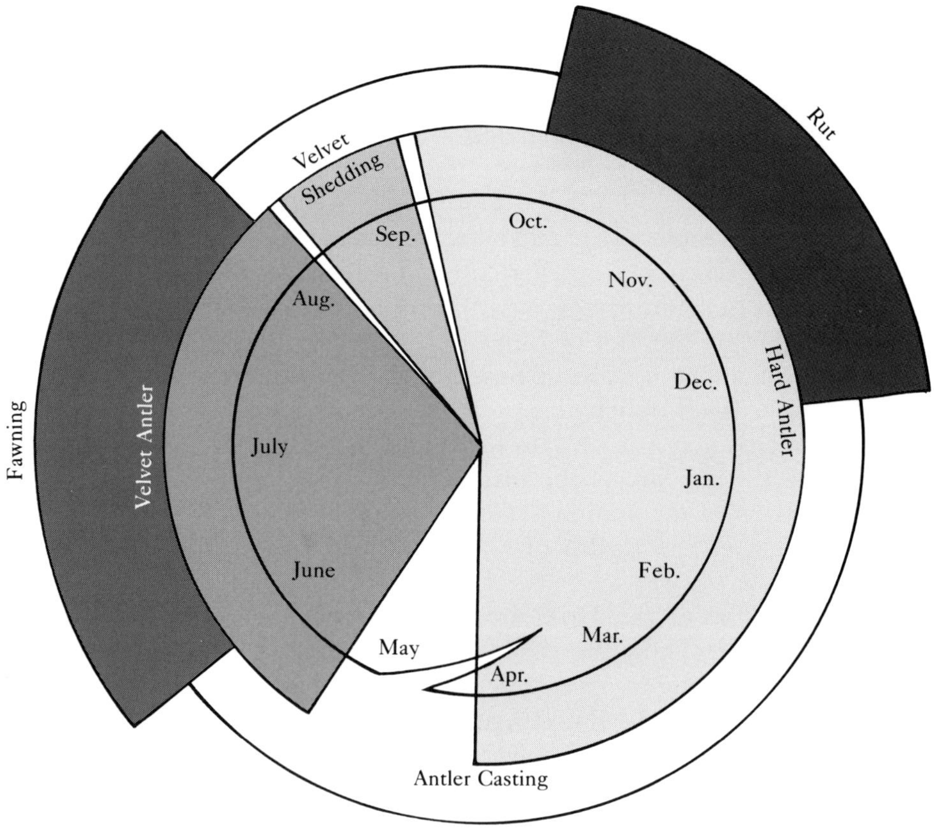

8

Behavior of Fallow Deer

The behavior of fallow deer is not easy to describe, because it is different from animal to animal, and also varies with the weather and many other factors. Fallow deer are social creatures. One finds them together in groups either grazing or resting. Does do not mind other fawns suckling, and I have seen as many as six fawns trying to suckle on the same Doe. It was not apparent which fawn was really hers, and chances are it was probably suckling another Doe as well.

Some deer are tame and stay that way throughout life no matter what happens. Some are tame on one day and spooky on another day. Some become tame with regular contact, and so on.

Most deer, however, get fairly tame if handled on a daily basis. Like most other animals, they respond to the person who takes care of them and feeds them every day. If you are willing to spend some time with them and give them a little treat every time you go behind the fence, you will be surprised how tame they'll become.

At one time, we had a small area of Lucky Star Ranch opened to the public. We had more than 6,000 visitors the first summer. Every morning at ten, when the first people entered the Deer Park, 80% of the deer came running to greet the visitors. The deer knew that they would be fed corn or crackers. Even the fawns, which normally are timid and flighty, came for their treat. The bucks, as long as they were in velvet, were also very tame. The only threat to visitors was the occasional stepped-on foot as the deer thronged forward begging for treats. Fallow deer are more timid and flighty than domesticated red deer. The most important thing to keep in mind, as far as the behavior of fallow deer is concerned, is that they can panic easily when handled inappropriately or pushed too hard. One has to be careful, and it takes a bit of experience to learn to handle them.

They are agile and can run fast. If they run, do not try to stop them. Once in motion, they would rather run you down than stop even if you stood in their way.

If crowded, fallow deer can jump high. I have seen deer jump an eight-foot fence without touching it. Normally, they would not even try to jump a five- or six-foot fence. They would rather try to slip underneath obstacles if they can find a hole. So make sure that your fences are tight and solid at the bottom.

If fallow deer stay in groups, one of the older Does will normally be the leader. When alarmed, however, they follow the very first animal which runs, even if this happens to be a fawn. One nervous animal can cause trouble even if the rest of the herd is hand-tame.

A significant trait of fallow deer is their bouncy gait, lifting all four legs together when running. The way they lift all four feet off the ground at once reminds me a little bit of the jumping of a gazelle. They can continue this gait for quite a long distance. One interesting thing is that fallow deer use this bouncy gait both when they are alarmed to the point of fleeing, and also when they are in an extremely good mood. I am normally pleased when I see my deer jump around like that. It is usually a sign of a happy atmosphere and, even if the deer have been alerted by something, as long as they are using their bouncy gait they are most probably not going to panic.

Fallow fawns like to run and jump in groups when they feel good. It reminds one of a group of children playing 'tag'. Other gaits the deer use are the walk, the trot, and the gallop. They also like to swim, which they do quite well, shaking their sides like dogs when they emerge.

If you bring a dog into the deer pasture, be careful, because Does may become belligerent toward dogs. If you want to round deer up with a dog, he has to be a good herd dog. An untrained dog will do more damage than good. Always keep the dog in front of the deer, because they tend to follow the dog in order to get him out of the pasture, and in that way, you can have the dog bring the deer where you want them. If you send the dog behind your herd as one would with sheep, you will find that the deer will all turn around to face the dog.

If one took the time to study the behavior of fallow deer carefully, one would probably do a better job than the weather-man in forecasting the weather for the next two or three days. If the deer are especially restless and flighty, there is probably a storm approaching. Sometimes in the fall or winter if the deer eat almost twice as much as on normal days, it is a sign of coming bad weather. If the appetite returns to normal, you know that the worst of the storm has passed.

During the summer months, the mature bucks tend to live apart from the Does. They do not move much, while they eat constantly to build up winter stores. The antlers grow and, by the end of August or early September, the deer start to rub off the velvet. During that time, they eat even more and prepare themselves for the rut season.

They need to gain condition and weight, because during the rut they hardly eat anything and lose up to 40 pounds. Eating habits of deer follow a daily rhythm. Deer normally start eating early in the morning for about two hours, including approximately 20 minutes of sleep. During the summer, until the rut, they have five to six eating periods with chewing periods interspersed. During the rut, this daily rhythm is disturbed. During the

Fallow deer are social creatures. They can become very tame and friendly.

Photo: Todd Bruno

winter months, deer normally only have two eating periods. (Van Acken)

According to Puttik, Does keep themselves busy during daylight with:

Grazing	27.50 %
Lying and Ruminating	7.21 %
Lying without Ruminating	5.31 %
Standing and Moving	14.40 %
Other	2.20 %
	56.60 %

accounting for 13.58 hours of daylight. The remaining 43.40% is figured for 10.42 hours of night. With bucks, those rhythms are not so clearly defined.

Deer eat with a quick and hasty biting of the grass, which is more ripped than bitten, as they press their cutting teeth against the dental pad. If the deer want to eat leaves or fruits of trees, they sometimes stand up on their hind legs, and can reach astonishingly high.

Fallow deer like to eat good grasses, legumes, all kinds of brush, branches of fruit trees, horse chestnuts, acorns, beets, carrots, potatoes, apples, oats, corn, barley, and wheat. They do need a high percentage of fiber, without which their ruminant system does not function correctly.

Feed should contain 11% protein, minimum, and during the winter months one has to provide approximately 65% of their diet as starches, or carbohydrates.

PLASTIC IS POISON!

For some reason fallow deer love to eat any kind of plastic, which is absolutely fatal to them. It is, therefore, important that you keep an eye open for any kind of plastic within your fences. Never use plastic baling twine.

Fallow deer do not like strong winds and driving cold rains, and in those conditions they do need shelter. Cold but dry weather, however, does not seem to bother them.

Signs of comfort are scratching with one hind leg, rubbing, licking the skin with the tongue, stretching, yawning, shaking, snorting, and wiggling the tail.

9

Purchase of Breeding Stock

When you purchase your first deer you can easily make your first mistakes. Condition of the animals should be carefully appraised. You do not have to be a deer expert to assess condition. How the deer walk will indicate if their feet are healthy and in good shape. You should not buy hoof problems, which themselves may be a sign of other systematic problems.

The skin of the animal should be evaluated. If the hair is slick and smooth and has a certain kind of bloom, the animal normally is in good health. The coat of a deer is the mirror of its condition.

If you buy deer from a game park, do not be mesmerized by the tame and cute animals. You want to farm the deer—tameness is an important trait, but it is not all that counts. Animals for a deer farm have to be healthy and well-framed. Your goal is to raise good healthy fawns and sell carcasses of at least 60 pounds. The first 50 deer we bought in 1978 came from Roland Lindemann of the Catskill Game Farm, in Catskill, NY. They were in good condition. The 300 deer we bought in 1979 were collected from all over the nation. Our aim was merely to build our herd as quickly as possible, since we thought then that a deer was a deer. This is not the case, and it has taken us years of breeding to finally establish a herd with good records, charting health and performance. We have raised our carcass weights from 35 pounds to 61 pounds and are now finally beginning to reap the harvest of our investment of eight years ago.

By buying functional, healthy animals from respected breeders, one can gain a few years and not necessarily pay more than at other places, where you may buy someone else's culls.

A West German study reveals that inbreeding with fallow deer reaches a "point of no return" after approximately fifty years. Until now, the last fallow deer imported into this country arrived about 1930, meaning breeders were probably approaching that point.

However, in 1986 Eliot and Susie Clarke of the Lithgow Deer Farm in Millbrook, New York, imported four breeding bucks from England, and in 1987 Lucky Star Ranch imported five breeding bucks from Sweden. It is our hope that the offspring of these two groups will bring good genes to all of us. One note: during the complicated and costly process of importing these bucks, we received great help and support from the USDA in Washington, D.C., for which we extend our thanks.

Never introduce new animals into your herd during the rut or during the fawning season.

10

The Farming of Fallow Deer

Fallow deer adapt readily to widely varying land topography, climatic conditions, and feed quality.

Optimum production, however, requires improved nutrition. The stocking and fencing investment involved in establishing a deer farm can be substantial, so one should, if possible, run deer on high quality or improved land to obtain maximum return.

That does not mean that deer need an expensive feed. But like any other animal, they do better if kept in good condition. Even if fed for superlative condition, not mere maintenance, deer will still give a better return than most other farm animals.

Fencing costs are also lower on easy ground than in heavily wooded areas or rocky terrain. Remember, the figure of four deer per acre only applies if that acre is good pastureland. It does not apply for woods or flat rock. When planning your pastures, please consider some shade for hot summer months. A wooded area or brushland is always handy to have within the pasture.

Investment:

Economically, starting a deer farm really only makes sense if you already have the land, the buildings, and basic farm equipment. In addition to these, you will still have to make the following investment:

100 Adult Does	$400. each	= $40,000.00
4 Adult bucks	$500. each	= $ 2,000.00
1 High quality buck (new bloodlines)		= $ 1,000.00
Fencing costs for a 25 acre pasture @$350./acre		= $ 8,750.00
Handling facilities, waterers etc.		= $ 8,250.00
Approximate investment		= $60,000.00

Variable annual costs:

Feed per animal, including fawns	$54.	= $5,670.00
Veterinary expenses, saltlicks, minerals, eartags, etc.	$ 4.50	= $ 472.50
Medication (adult and offspring)	$ 2.00	= $ 210.00
Miscellaneous		$ 147.50
		$6,500.00

First return after two years:

100 females—90 yearlings	@ 300.	$27,000.00
(female yearlings could bring more, possibly	@ 350).	
Subtract costs for two years	@ 6,500./year	$13,000.00
		$14,000.00

NOTE: Remember that you have borne the cost of two years before your first sale, so each additional year could yield, for example, $20,500 instead of $14,000.

OR:

Keep the females to build up the herd, and sell:

45 male yearlings	@ $300./ ea.	$13,500.00
Subtract costs for two years	@ $6,500./ year	$13,000.00
		$ 500.00

In this case, the profit is nothing to write home about, but don't forget, you have built up your equity by . . .	$13,500.00
so the result, in the long run, will be the same.	$14,000.00

As with all other farm animal operations, the success of deer farming depends largely upon the knowledge and the abilities of each individual deer farmer.

The market production of each animal can vary from 35 to 65 pounds carcass weight. The price per pound can fluctuate between $3.00 and $5.00. In individual cases a greater premium may be paid.

A 35-pound carcass (the average weight at the Lucky Star Ranch in 1980) at $ 4.00/pound = $140.00

A 61-pound carcass (the average weight at the Lucky Star Ranch in 1985) at $ 4.00/pound = $244.00.

The 61-pound average was the average out of a group of 150 deer sold in 1985. A 61-pound carcass (which one day can hopefully be raised to 65 or 70 pounds) sold at our current price of $5.00/pound = $ 305.00.

There may easily be a difference of $100.00 or more per animal sold; that can make or break a farmer. Success requires hard work and serious efforts in this industry, as it does in any other, but the potential here is vast.

If you can, try to sell some of your fawns, especially females, as breeding stock. For a breeding female, you can ask almost the same price at the age of six months as you would ask for a yearling spikehorn at 15 to 18 months. This will save you the cost of feeding, plus all the marketing and slaughter expenses. Just be sure to sell only top quality females; otherwise, you cannot build a good reputation and a long-term business. The old mentality that one did well cutting corners just short of cheating one's neighbor needs changing. Only people who are honest and have integrity are going to survive in the business over time. If you serve your customer in a better way than others, give him better service and better quality, you have laid the foundation for success.

If things go well, a 90 percent fawn crop in a herd of adult females can be the rule. If you start out with a group of young females, the fawn crop might not be greater than 70 percent for the first year—perhaps approaching about 80 percent the second year.

If you drop below 80 percent in an adult herd, you have to look for the reason. Look for predators like coyotes or vultures; we have had both at the Ranch. We carefully check the fences two or three times a week. Human scent at the fence lines repels the coyotes for a few days. You can also put an electric wire around your fences or on top of the fence, because coyotes are supposed to be able to climb fences.

Deficiency of trace minerals can be a big problem. Have your veterinarian help you with this problem because it is tricky; too much selenium, for example, can be just as fatal as too little. Regular forage sampling and analysis combined with blood testing can help you stay on top of things.

If you cannot keep a minimum of three animals per acre of supposedly good pasture, you have to improve your pastures. Four animals per acre is the rule. If your pastures are in top shape, you may be successful with five adult animals per acre.

If you plan to make deer farming more than a hobby, you should not have fewer than 50 Does. Otherwise the fixed costs would be too high, making it unprofitable.

The three most important factors are:

Sale price
Carcass weight
Fawn crop percentage

Optimizing those three factors will be the key to success. Another important factor will be the cost of feeding your animals and raising fawns.

According to Julie McCall, from New Zealand's Farm Advisory Office, in her study "Effects of Performance and Prices on Fallow Deer Profitability," the most significant increase in fallow deer profitability can be made through increasing fawning percentage. This holds for both low and high Doe and venison prices, although the importance of fawning increases as Doe price increases.

11

Pastures

The size of your pastures depends on your specific situation. I think it should not be less than four acres, because that would result in a poor return on your investment. A 25-acre pasture would be preferable, with a subdivision for rotation, if needed. One hundred deer would be a desirable operation, economically, even though 50 deer can, in most cases, provide a nice little income.

The design and layout of your fences will determine how difficult or how easy it will be to run the entire operation. The ideal situation would be to have a solid piece of some 100 acres with no fences or buildings that need to be utilized. One could build a system of fences, combined with work areas and shelter, which, for example, would look like the one on the following page.

The layout has an important bearing on the workability of a deer farm and its management. That is why you should spend time looking at other operations and carefully planning. Find out where other people have made mistakes and avoid making them yourself. If you come to the Lucky Star Ranch, I can show you hundreds of mistakes we made and had to correct. When we began, we didn't have anyone to turn to for advice. You do! Feel free to take advantage of your lucky situation.

It is important that the manager or owner of the farm live on the property. It will keep predators away, four-legged and two-legged ones. Besides, if the deer are familiar with the people living on the place and know the vehicles that come and go, they tend to be quieter and easier to handle.

Pasture access for vehicles is quite important and should be considered from the beginning. Do not forget to allow access for trucks that transport the deer.

Within the fences there must be good shade for hot summer days. Fawns seek cover and protection during their first few weeks of life. This must be available to them. A good way of providing cover is to leave an area in the middle of the pasture relatively long and unclipped, so that the fawns can lie down in it. This keeps them away from fences and will prevent the fawns from slipping through the fence in order to find cover outside. They might not make it back through the fence and may either starve to death or

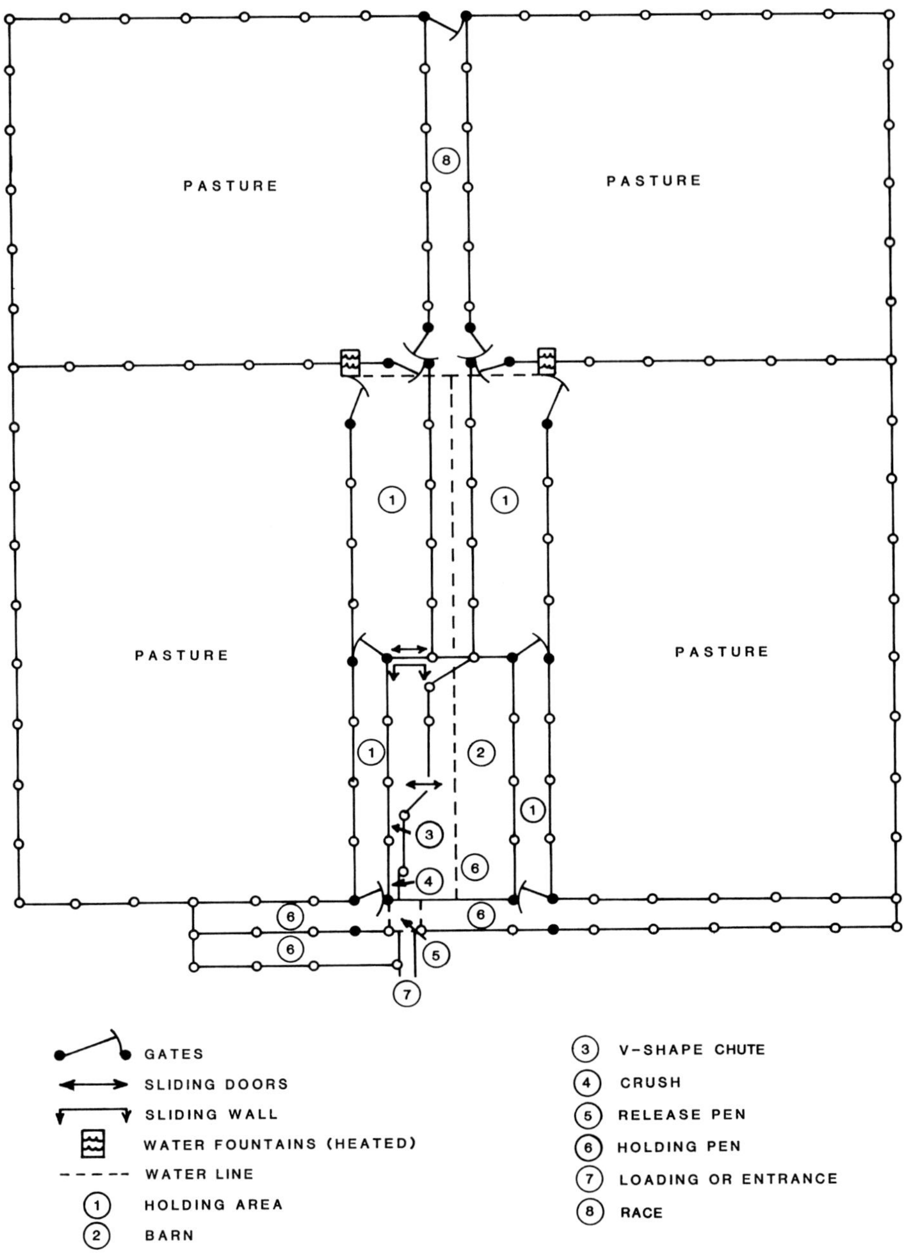

be taken by predators.

All pastures must have sources of good, clean drinking water. Fallow deer are supposed to drink at the very least a quart per day. During winter months, it is advisable to use heated drinking elements, like Ritchie Fountains.

The ideal pasture layout should have a raceway in the middle, so that you can herd your deer from any pasture to your working area separately. This raceway should taper to a width of six to nine feet at the entrance to

A drab, old, white Doe became very dear to us, because she will come whenever we call her and lead the way for the others. We named her "Chrissy". *Photo: Author*

the work yards. There should be several gates along the race, so that you can close the passage as soon as the deer have gone past them. This prevents the deer from running back all the way.

To make the herding of your deer as easy as possible, the gates must be planned carefully. In my opinion, gates in pastures should always be located in corners and not in the middle of the fence line. Deer have a tendency to run into the corner and go past the gates in the fence lines. After we found that out, we changed all our gates to the corners, a costly project.

If it is impossible to have one raceway with direct access to all pastures, try to have a race from a collecting paddock and bring the deer there first.

Whichever way you work it out and whatever the contours of the land or other existing factors may be, try to come as close to the ideal as possible. It is important that the deer are familiar with the area and especially with the position of gates. It can take a few days until the deer are willing to walk through a new gate with which they do not feel familiar. A drab, old, white Doe became very dear to us, because she will come whenever we call her. She will lead the way for all the others if we want the herd of Does or the fawns to go through a certain gate into another pasture. We named her "Chrissy," and even though she must be more than 15-years old, she still has a beautiful fawn every year.

If your farm is located in rolling hill country, keep in mind that deer generally will run uphill more readily than downhill. Also, provide extra pastures for the breeding herd and separate them from the young stock.

Fallow deer are a little more difficult to yard and handle than red deer, because they are nervous and can easily excite to the point of panic. If you know what you want from them and know how to handle them, your job becomes easier. Learn their methods of escape and then be smarter, quicker, and better prepared the next time. If you reach a point at which you have to herd them up over and over again and they still get away from you, Stop! Think! It is time to make some changes because something is definitely wrong. You have to find out what it is before the deer do. Do not shout and do not chase them. Let them calm down. And if it takes them one-half hour to walk through a gate, do not push them, give them time. It is easier just to stand there and wait, then to go back and bring them around again and again. Lead fences or wing fences are very helpful if they have the right height. A four-foot snow fence or page wire will not stop them. They will jump over low fences when crowded.

The good deer farmer will bait his deer single-handedly into any pasture, raceway or yard, with a little grain or some crackers, or apples. He will not need any help until the deer are actually in the working area to be handled.

Pastures should be dry and in good condition. (The name fallow deer does not indicate that those deer can live off fallow land.) Deer can certainly survive on maintenance feed, but if a farmer expects a good return on his investment, he should do everything possible to ensure superior results. A dairy cow that gets only hay of inferior quality cannot perform as well as one that gets high protein alfalfa. Basically, the same applies to deer, although they need a lot less to perform than a cow.

If a pasture would normally support 1 cow per 1½ acres, that same pasture would be adequate for 4 production units fallow deer (PUF) per acre.

1 production unit fallow deer (PUF) consists of:

1 adult Doe—0.85 fawn—0.05 buck

The bucks are counted in an average ratio of 1 buck for 20 Does. This ratio can vary according to how many Does run in one herd together.

Pastures should be dry and in good condition. *Photo: Joseph Albino*

One example of ideal pasture make-up might be:

8%	Dutch white clover	(*Trifolium repens*)
8%	Bird's-foot trefoil	(*Lotus corniculatus*)
4%	Hop trefoil	(*Trifolium campestre*)
15%	Orchard grass	(*Dactylis glomerata*)
9%	Red Fescue	(*Festuca rubra*)
20%	Timothy	(*Phleum pratense*)
8%	Kentucky bluegrass (June grass)	(*Poa pratensis*)
4%	Greater burnet saxifraga	(*Pimpinella major*)
6%	Narrow leaved plantain (Rib grass)	(*Plantago lanceolata*)
1%	Wild parsley	(*Petroselinum sativum*)
7%	Alfalfa	(*Medicago sativa*)
10%	Blue grass	(*Sesleria coerulea*)

(Fritz-Berthold Bamberg)

It is not absolutely necessary, however, to be so specific. Any *good* pasture will do. Have regular soil samples taken and fertilize accordingly. A recommendation from Germany, converted into pounds per acre, would be:

70 to 90 lbs. P_2O_5, in the form of Thomasphosphate.
100 lbs. K_2O
50 to 60 lbs. MgO (Magnesium)
200 lbs. Na_2O (Natrium)

It is advisable to include some brushland in the pastures.

Photo: Author

Nitrogen should be used as needed.

It is quite helpful to rotate the animals, if possible. Keep free choice salt available at all times and if necessary supplement minerals and selenium to prevent deficiencies.

It is advisable to include some brushland in the pastures. Fallow deer require a high fiber diet and brush and branches are like medicine to them. Deer love apples, so apple trees are of value in pastureland. They also love the branches of any kind of fruit tree, so if you are pruning your trees, don't throw the branches away. Give them to the deer.

Because deer seek a high fiber diet, they will eat the bark of smaller trees, thus killing the trees. In order to prevent this, European foresters have come up with a surprising method. They put out boxes that are filled with natural peat moss as a kind of free choice feeder for wild deer. The peat moss supposedly contains some of the same acids that deer are looking for in the bark and limbs, and also contains a large amount of fiber. At Lucky Star Ranch we are in the process of trying out this method, and we are looking forward to the results.

Be careful during fawning time. Do not push Does into a new pasture and forget about the little fawns in the old one. If you separate the pastures internally with only one or two electric wires, it will allow any remaining fawns to slip under the wires while preventing the Does from crossing back into the old pasture. It is not a big problem if one or two Does happen to slip through as well.

Here are some ideas for good hay or pasture mixes:

For hay and pasture on more acid, poorly drained soil:

60% Red clover
20% Timothy
20% Alsike clover

For pasture or marshy field:

45% Alfalfa
25% Tall fescue
20% Orchard grass
10% Timothy

High yield for haylage:

60% Alfalfa
15% Red clover
15% Timothy
10% Alsike clover

Good hay mix for young stock and weaned fawns:

70% Alfalfa
15% Red clover
15% Timothy

Legume/grass mix for multiple use:

20% Fescue
20% Alfalfa
20% Kentucky bluegrass
15% Orchard grass
15% Red clover
10% Timothy

12

Fencing

Fencing can be one of the largest expenses when starting a deer farm. Improperly selected and installed, this area can cost you more money in one year than you could make in years of successful deer farming.

When we started our deer farm at Lucky Star Ranch, in 1978, we were told that exotic animals had to be fenced behind a seven-foot chain link. That piece of incorrect information led us to a costly mistake, with the expensive purchase of our first chain link at $2.75 a foot, plus galvanized posts and hardware and other equipment. When our first fallow deer arrived, we had fenced a ten acre pasture and paid much more than we would have to pay today for fencing around a 25-acre pasture.

Today we know better, and we now work exclusively with a special deer fence from New Zealand, called "Tightlock". Tightlock high-tensile deer fencing is used around the world for deer control. Originally developed more than 10 years ago in New Zealand, it is now used by American deer farmers as well as for the protection of horticultural sites from native deer. The development of this fence was a real landmark in the realm of products specially designed for the needs of deer farmers.

The high-tensile, woven wire is 75 inches high and is topped with smooth wires. At Lucky Star we have two strands of electric wires on top to keep out four-legged and two-legged predators. Tightlock fence stays tight without stretching and sagging like softer wire, and is galvanized to last many years. After a decade of experience, we can truly say this: to our knowledge it is the best deer fence available and is lower in price than any other even remotely suitable fencing.

The Tightlock fence originally had 13 line wires, and we had some problems with fawns slipping through. Since then we have convinced the manufacturer in New Zealand, with whom we work closely, to add one more horizontal wire, giving it 14 line wires, spaced 3½ inches apart toward the bottom. We believe this has solved the problem of fawns escaping.

The stay wires, or vertical wires, are six inches apart, and the rolls are 20 rods long (330 feet). The Tightlock construction has tight grip knots and stiff, one-piece vertical wires. The fencing should be double braced in each direction and posted every 20 to 25 feet. To make a tight and springy fence, the wire is stretched to flatten the line wire tension crimps half-way (stretched about 5 feet per roll). These crimps help to keep the tension, allow one to fence over uneven ground, and add to the fence's impact resistance. Bracing and dip posts should be four or more feet deep in the ground, and firmly set. Brace rails should be 10 feet long and brace wired in place.

If some of this sounds a bit like a commercial for "Cyclone Tightlock" fencing, it is intentional, because I am convinced that there is no better fence for the same price available, and I highly recommend using it. This kind of "full netting" fence may seem expensive compared to electric wire fencing, but it lets a deer farmer sleep a lot better, since, according to our experience, fallow deer don't seem to be deterred much by electric wire fences.

On the following pages we will explain how to build a solid deer fence. The Cyclone company gives detailed instructions for the installation of their fencing, and we have made a few modifications, which are described here. You can combine our methods with their instructions according to your own requirements. There is also a video tape available that shows, in detail, how to build deer fences, and which contains, in addition, much else that would be of interest to deer farmers. This video can be purchased from the Lucky Star Ranch Corp., Chaumont, NY 13622. Telephone (315) 649-5519.

Here is what we do at the Lucky Star Ranch:

1. Clear brush, making a path about 8′ wide.
2. To set the main corner post, auger or dig a hole about 4′ deep. Clean out the dirt, set the post and pack filling of alternate shovels of dirt and gravel, say, three or four of dirt, then gravel, etc. Tamp it down tightly.
3. Run a stringer wire from corner post to corner post, making sure stringer is on the inside of the posts.
4. After putting up the stringer, measure 10′ from the corner and set the 1st vertical brace post, using same procedure as corner post.
5. Repeat this for other brace posts—(two on each side of corner post)
6. Run a diagonal brace post from bottom of 1st vertical brace post to main corner post, joining approximately 5′ from the ground. Do the same thing between 1st and 2nd vertical braces.
7. Put up horizontal braces, approximately 5½ feet from the ground
    a) between 1st vertical and corner
    b) between 1st vertical and 2nd vertical

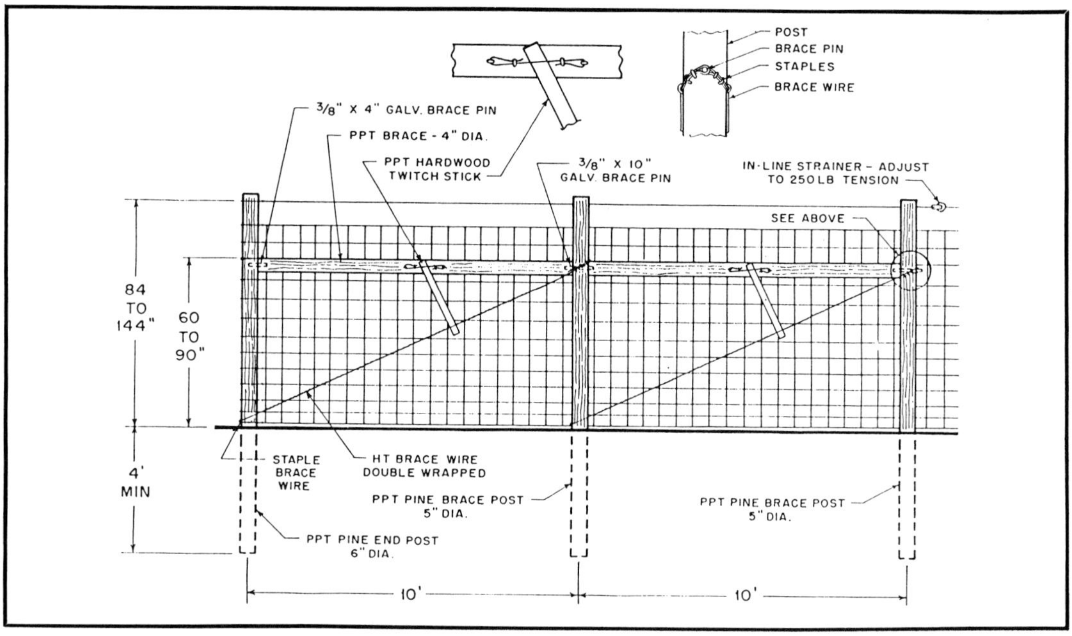

Double Horizontal Brace Assembly

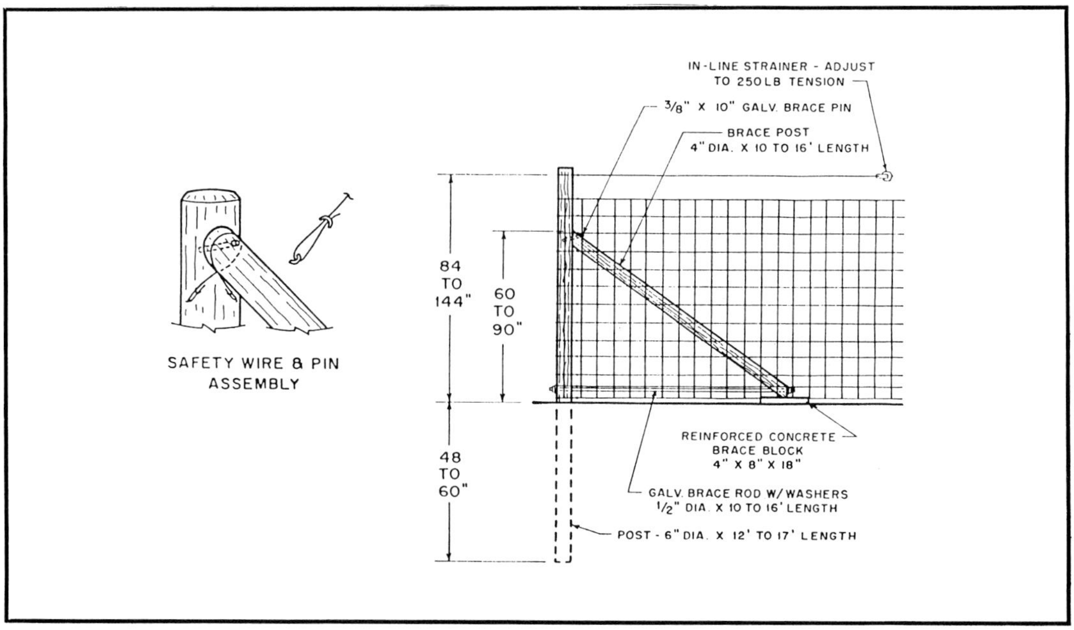

Diagonal Brace Assembly

Brace Assemblies:

Courtesy of Cooperative Extension Service, West Virginia University, (Publication 812)

8. Put diagonal brace wires on corner post, crossing diagonal brace posts.
    a) Using 12-gauge soft wire, wrap around bottom of main corner and cross to 1st vertical above the horizontal brace and back again twice.
    b) Do the same between 1st vertical and 2nd.
    c) Tighten brace wires by inserting a wood batten in the wire loop and twisting it toward the outside until wire is snug. When it tries to twist back, hook the top of the batten outside the horizontal brace to hold it.
    d) Do the same for each brace.

9. Line posts should, ideally, be set about 8 paces apart (20′–25′) on straight stretches. On contoured terrain, however, posts must be set at each deviation, whether a depression or rise. Only in this way can the woven fence be "tailored" to the terrain, thereby reducing the labor required to secure the bottom of the fence. Line posts in depressions must be "footed" to counteract upward thrust when the woven wire is stretched.
   Pace off the distance between the 2nd vertical brace on corner to the other corner on that line, and divide the total paces by 8 to give you the number of line posts. Set markers about 3 inches behind the stringer wire.

10. Set line posts the same as corner posts.

11. When you trim the post tops at 8′ or 9′, cut at an angle so the rain runs off. This prevents early rotting.

12. Pull stringer back from the post line to allow room for the laying-out of the woven wire.

13. When you lay out the woven wire:
    a) make sure the woven wire is on the inside of the posts,
    b) make sure the smaller spacings are snug against the bottom of the posts,
    c) roll out the wire allowing about 2′ to wrap around corner posts,
    d) tie wire around corner posts using splicing tool
    e) where standing trees are used as end or corner posts, allow 1′ in order to tie off woven wire to "J" bolts
    f) do the same on the other corner

14. Pull the woven wire tight, pulling from the corners to the middle of the fence.

15. Stretch woven wire tight by using the strainer boards provided by the fence company.

16. Staple fence to line posts but do not drive staples home. The fence has to be able to move.

Corner post with two vertical brace posts on each side, 10 feet apart from each other, and diagonal brace posts, joining approximately 5 feet from the ground.

Photos: Author

Double Horizontal Brace Assembly *Photo: Author*

If the ground is too shallow to auger 4 feet holes, we use this method of building legs and holding the post down with heavy rocks.

Photo: Author

FENCING COSTS

Example from Lucky Star Ranch, Chaumont, NY 13622

Area Fenced: Approximately 23.25 acres
Perimeter: 4,180 feet
To determine how many rolls of fence:
Divide 4,180 feet (perimeter) by 330 (length of fence roll) = 12.66 or about 13 rolls of fence

4,180 feet = 13 rolls @ $.70/ft or $231.00/roll	= $3,003.00
Posts at approximately 20 feet apart	
= approx. 290 posts @ $ 3.00/ea.	= $ 870.00
Bolts, staples etc.	= $ 584.00
Materials	= $4,457.00
Labor 30 days (8 hour) @ $10.00/hr.	= $2,400.00
Stretching tools and other equip.	= $1,200.00
	= $8,057.00

$8,057.00 for 23.25 acres = approximately $ 346.50/acre

Deer prefer to remain outside even during sub-zero temperatures. As soon as it becomes wet, however, they all crowd into the barn.

Photo: Irmel van Meeteren

13

Shelter

Most visitors to Lucky Star Ranch ask me if the deer really need shelter, since they are still fairly wild animals. In a properly managed and well hunted forest, there would be perhaps about ten deer living on 250 acres and they would definitely find enough natural shelter. In our situation, however, the circumstances are different, with a population density that we would never find in the woods. Therefore, we have to provide shelter for them during the cold, wet season, when we can expect freezing rain and generally harsh weather.

Our deer barn at Lucky Star has a concrete floor, an adjoining silo with an unloader, and an automatic conveyor-belt feeder, not all of which is really necessary. For the most part, our deer ignore it, preferring to remain outside even in sub-zero temperatures. As soon as it becomes cold and wet, however, they all crowd into the barn.

It is important that the shelter is not a dark, damp place, which could

A plastic greenhouse used as a deer barn in Germany, has been fenced off along the outside, so that the deer cannot reach the plastic.

become a breeding ground for all kinds of diseases. Constant fresh air going through a dry barn is important.

A shelter for deer need not be fancy or expensive. Anything from an old barn to a simple lean-to will do. A pole barn enclosed on three sides is perfectly adequate. One of the best ideas I have seen is a plastic greenhouse used as a deer barn on the farm of Josef Rüssmann in Germany. It makes a perfect shelter and is not at all expensive. As you can see in the photos, the plastic greenhouse has been fenced off along the outside, so that the deer can't get to the plastic. Along the inside are hay feeders, creep feeders, etc. Some plywood to keep the deer from the plastic walls would be a good idea.

If adequate, dry shelter is not provided during cold and wet winters, the animals become severely stressed. This stress results in a decrease in fat reserves. This, in turn, increases the problem, because skin fat has the function of acting as an insulator in cold weather, protecting the animals from hypothermia. Since fat reserves during wintertime cannot be increased, young animals will very likely be in bad shape by January.

In the summer, shelter is also required if there are not enough brush or wooded areas within the fences. Deer need shade from direct, hot sunlight.

Along the inside are creep feeders, hay feeders etcetera. *Photos: Author*

The german deer farmer Josef Rüssmann and his son Georg in the plastic greenhouse which they converted into a deer barn. On the left side a creep feeder for fawns.

Photo: Author

A shelter for deer need not be fancy or expensive. Anything from an old barn to a simple lean-to will do. A pole barn or a movable hut is adequate.

Photo: Author

Feeding off the ground is not absolutely necessary but, if possible, advisable. Troughs should be lower on one side and open to allow the water to run off. They can be used for haylage, silage, grain etcetera. Our troughs are appr. 22 inches high at the higher side and 20 inches at the open side.

Photo: Author

14

Feeding

A clear understanding of the feed requirements of farm animals will not only result in increased production and healthy stock, it will also assist greatly in farm planning and budget management. In this chapter, we will outline the feed requirements of fallow deer based on typical circumstances in the four seasonal periods, and the patterns of weight gain and weight loss.

But first, a few words of good common sense:

"The eye of the master
fattens the beast!"

That is how my good friend Wib Donaldson from Canada put it, and he adds: "As important as WHAT you feed your animals is HOW you feed them." He is absolutely right. One can spend a lot of money on grain and expensive pellets, but thrown on the ground or dumped into a dirty trough to turn sour or collect bird droppings, the animals will hardly touch the food and you are wasting money. On the other hand, if you take good but reasonably priced grain and feed it in a clean container, your animals will finish it all. A little Tender Loving Care goes a long way, and it doesn't cost anything.

Feeding requirements differ in each of the four seasons. There is a general pattern of liveweight gain and loss around which feeding patterns must be planned.

Young fawns show a low rate of gain over the autumn/winter period following weaning, at 100–120 days of age, with a high rate of gain during spring/summer when the animals are 9–15 months old. During this period, the liveweights of most deer will roughly double, from approximately 40–50 pounds to approximately 90–110 pounds on average.

Older bucks have an unavoidable weight loss during the rutting season in autumn, followed by a further weight loss in winter. High levels of feeding during this period can minimize the winter weight loss, and we recommend this highly, since groups of animals under nutritional stress are more susceptible to disease outbreaks.

There may be a slight weight gain among adult Does in late pregnancy and during lactation, but not usually enough, we have found, to be significant.

BUDGETING FOR FEED PER HEAD

This brings us to the question of what to feed, how much, and when. The first consideration is feed quality. The energy in food that is actually available to the animal is called metabolisable energy (ME). This food value is expressed as a percentage of the whole bulk, and for comparison, we list the ME percentages of selected feeds below.

FEED	%Dry Matter	ME%
Ryegrass/white clover pasture		
Autumn	15	10.8
Winter	15	11.2
Spring-short	15	12.0
Spring-mixed	15	11.2
Summer	18	10.3
Meadow Hay		
Young leafy	85	9.0
Mature	85	8.0
Weathered	85	7.0
Grains		
Barley	85	12.5
Wheat	85	12.5
Oats	85	11.5

As presented in Ulyatt *et al.*, MAF/NZ (1980)

As one can see, spring pasture and grains provide the highest feed quality, with hay as supplement at various lower values.

An important variable in the condition of fallow deer is the development of the kidney fat capsule. This fat capsule can only develop during the late summer and early autumn. With a kidney fat index devised by T. Riney (1955), K. Baker used kidney fat capsule measurements to check the condition of fallow deer in the Blue Mountains on the South Island of New Zealand in 1973. Only a good fat reserve enables the animals to survive severe winter conditions, during which the metabolism converts to fat-burning. This cycle is maintained even after years of living in enclosures and under optimum feeding conditions. Inefficient feeding in autumn cannot be compensated for by optimum feeding conditions in the winter months. Young animals, improperly fed during autumn, will have used up their fat reserves by January and will most likely die. Most farm animals, during their long period of domesticatiön, lose this cycle of building and using fat reserves. Deer are not yet conditioned enough to ignore their natural cycle. So, in order to prevent heavy losses, we have to consider feeding our young animals grain, beginning August 1st, by using the creep-feeders shown in this book. The capsule of fat around the kidney, also known as the "*capsula adiposa*," should weigh about twice as much as the organ itself when fully developed.

With good body fat reserves, fallow deer will breed earlier in the season, and the breeding will be concentrated in a shorter time span. Animals in good condition will conceive early in the breeding cycle, and generally have a higher rate of conception.

During winter, there are many different combinations of feeding. Always keep in mind that you should budget for approximately 2.2 pounds of dry matter per 88 pounds (40 kilograms) of body weight, and that this feed should contain at least 11% protein and approximately 65% carbohydrates.

During January and February, you can cut down on your feeding costs by using corn silage instead of grain, as long as you provide enough energy food and enough grain to carry the minerals, especially selenium.

Fawns should get grain during all of the winter months.

A test in Bavaria, West Germany, conducted by Prof. Dr. Hermann Bogner and U. Zeeb and K. Popp, at the Bayerische Landesanstalt für Tierzucht, shows that 1 adult fallow buck, 13 adult Does, 5 yearling Does and 9 fawns consumed 1,914 pounds of hay, 5,720 of haylage and 726 pounds of grain during a period of 100 winter days.

First, we figure their average body weights at:

1 buck	@ 145 lbs.	145 lbs.
13 adult Does	@ 110 lbs each	1,430 lbs.
5 yearlings	@ 75 lbs. each	375 lbs.
9 fawns	@ 50 lbs. each	450 lbs.
		2,400 lbs.

Then, using the weight of 88 pounds as a rough average animal body weight for this group, we see that they consumed 2.06 pounds of dry matter per day per 88 pounds of body weight, which seems to prove our rule of thumb of feeding approximately 2.2 pounds of dry matter per 88 pounds of body weight.

If we wanted to see what an average adult Doe might consume, we take the figure 110 pounds for one adult Doe and divide it into the entire amount consumed (2,400 pounds/100 days) and get an average measuring group of "22 animals," which consumed:

19.14 lbs. of hay/day	= 0.87 lbs. per animal/day
57.20 lbs. of haylage/day	= 2.60 lbs. per animal/day
7.26 lbs. of grain/day	= 0.33 lbs. per animal/day

Translating this into:

2.6 lbs. of haylage	= 1.56 lbs. of dry matter
.87 lbs. of hay	= 0.74 lbs. of dry matter
.33 lbs. of grain	= 0.28 lbs. of dry matter
	= 2.58 lbs. of dry matter/per 110 pounds.
Or	= 2.06 lbs. of dry matter/per 88 pounds.

FEED COSTS

GRAIN		
@ $200. per ton	1 lb. of grain will cost	$.10
$160. per ton	1 lb. of grain will cost	$.08
$140. per ton	1 lb. of grain will cost	$.07
$120. per ton	1 lb. of grain will cost	$.06
$100. per ton	1 lb. of grain will cost	$.05
SILAGE		
@ $ 30. per ton	1 lb. of silage will cost	$.015
HAY		
@ $100. per ton	1 lb. of hay will cost	$.05
$ 80. per ton	1 lb. of hay will cost	$.04
$ 75. per ton	1 lb. of hay will cost	$.0375

Each individual deer farmer can assess his feed budget for the year and calculate the cost of various feed combinations to find the most economical system of supplementary feeding.

If there is plenty of pasture there is no need to feed the deer during the summer months, except for minerals and the salt. Don't forget, however, to creep-feed the fawns beginning August 1st. Since fawns have to learn from their mothers how to feed out of a trough, the troughs should be placed inside the creeps within reach of the Does, so they can stick their heads in and teach the fawns. Later, the trough can be placed farther into the creep so only the fawns can reach it when they enter. The idea of summer feeding of fawns comes from Josef Rüssmann. He also believes that creep feeders have to be in place before fawning begins, otherwise the fawns might not use them.

Water trough with small-animal-side-drinker

Photo: Author

For winter feeding, there are many possibilities. Here are two examples that we have used with success at the Lucky Star Ranch:

Oct–Dec	1 lb. grain/day animal	($140.00/ton)	=	$ 6.30
	2 lbs. hay/day/animal	($ 80.00/ton)	=	$ 7.20
Jan–Feb	½ lb. grain/day/animal	($140.00/ton)	=	$ 2.10
	2 lbs. corn silage/day/animal	($ 30.00/ton)	=	$ 1.80
	1 lb. hay/day/animal	($ 80.00/ton)	=	$ 2.40
Mar–Apr	1 lb. grain/day/animal	($140.00/ton)	=	$ 4.20
	2 lbs. corn silage/day/animal	($ 30.00/ton)	=	$ 1.80
	1 lb. hay/day/animal	($ 80.00/ton)	=	$ 2.40
				$28.20

OR

Oct–Apr	1 lb. grain/day/animal	($140.00/ton)	=	$14.70
	2 lbs. corn silage/day/animal	($ 30.00/ton)	=	$ 6.30
	1 lb. hay/day/animal	($ 80.00/ton)	=	$ 8.40
				$29.40

Feed costs fluctuate substantially with changing feed prices.

NOTE! During the winter months, it is important to remember these two objectives: 1) priority feeding to young stock, and 2) maintaining Doe live-weights and conceptus growth. Watch for aborted fetuses and have the animal tested for disease. If possible, mark the Doe for future testing.

As yet, no period of pregnancy has been identified as being particularly sensitive nutritionally in determining birth weights, but be cautious about any period of severe undernutrition during pregnancy.

WATER

Fallow deer require an average of 1–3 quarts of water per day. During cold weather, don't forget to check the water fountain heating system to make sure the water isn't frozen.

MINERALS

The following is a list of minerals known to be required in supplementation for fallow deer:

1. Selenium
2. Magnesium
3. Iodine
4. Iron
5. Calcium
6. Phosphorus
7. Zinc
8. Manganese
9. Nitrogen
10. Potassium
11. Sulphur
12. Cobalt
13. Copper
14. Salt

Trace minerals and the effects of mineral deficiencies are areas where man's current knowledge is incomplete, even to the extent of knowing minimum

Weather protection for salt and/or mineral container *Photo: Author*

mineral levels. From what we have found, our part of the country is deficient in selenium and deer need selenium supplementation. An overdose, however, can be fatal. Always work together with your veterinarian.

The mineral mix we use is called Fortimin, and is available at feed stores. It contains:

Phosphorus	10%	Iron	280 ppm
Calcium	19 to 23%	Manganese	1600 ppm
Magnesium	3%	Zinc	4000 ppm
Sulphur	0.75%	Selenium	20 ppm
Potassium	0.6%	Vitamin A	300,000 USP units/lb.
Iodine	120 ppm	Vitamin D_3	150,000 IU/lb.
Cobalt	30 ppm	Vitamin E	300 IU/lb.
Copper	200 ppm		

Deficiencies in minerals can cause problems. Here are some examples:

IF YOU HAVE:	ELEMENT NEEDED: (in soil or supplemented)
Mastitis or lame animals	Magnesium
Scours in fawns, pneumonia	Magnesium
Animals die suddenly	Magnesium
Grass tetany on early pasture	Magnesium
Bloated animals on legume pasture	Potassium
Mange and ringworm	Boron
Warts on livestock	Iodine
Animals eating wood, bark, etc.	Manganese
Cystic ovaries, repeat breeding	Manganese
Hoof rot	Copper or Iodine
Slow growth of young animals	Zinc
Abnormal bone formation, embryo	Zinc

COMMON PASTURE PROBLEMS

DEFICIENCIES IN CROPS	CAUSE
Yellow top on second crop of alfalfa, bitter flavored	Lack of boron
Nitrates in plant tissues, slow growth, bitter taste	Lack of boron, sulphur or molydenum
Legumes with black spots	Lack of magnesium
Legumes with white spots on leaves	Lack of potassium
Grass leaves dead on tips and edges	Lack of potassium

This chart shows the feed contents as far as dry matter, raw protein, energy, calcium and phosphorus are concerned, based on tests in Germany and our experience. Always keep in mind the figure of 2.2 pounds (1 kilogram) of dry matter per 40 kilogram (88 pounds) of body weight per day and also the figure of 11% protein and 65% energy (winter feed).

Hay & Straw	% dry matter	% protein	% energy	Ca	Ph
Meadow hay					
1. cutting	87	9.7	33.6	6.3	2.3
2. cutting	86	11.0	28.9	8.2	2.5
Straw					
Wheat straw	89	2.6	23.2	2.8	0.7
Barley straw	87	2.9	30.0	2.6	0.7
Oat straw	88	2.9	28.9	3.6	1.2
Silage					
haylage 1. cutting	30	3.9	16.5	2.2	1.0
2. cutting	44	5.6	22.2	3.2	1.5
Grain					
Barley	87	10.2	71.9	0.6	3.6
Oats	88	11.0	64.5	1.0	3.1
Wheat	87	11.5	75.1	0.6	3.3
Corn	87	9.4	74.6	0.3	2.8
Brewers' fresh	23	5.8	13.6	0.9	1.6
silage	27	6.7	13.9	0.9	1.5
dried	92	23.2	42.3	4.1	6.6
Sugar beet pulp fresh	13	1.1	9.0	1.3	0.2
dried	90	9.0	63.0	9.1	0.9
dried with molasses 16–22% sugar	90	9.0	61.0	7.5	0.7

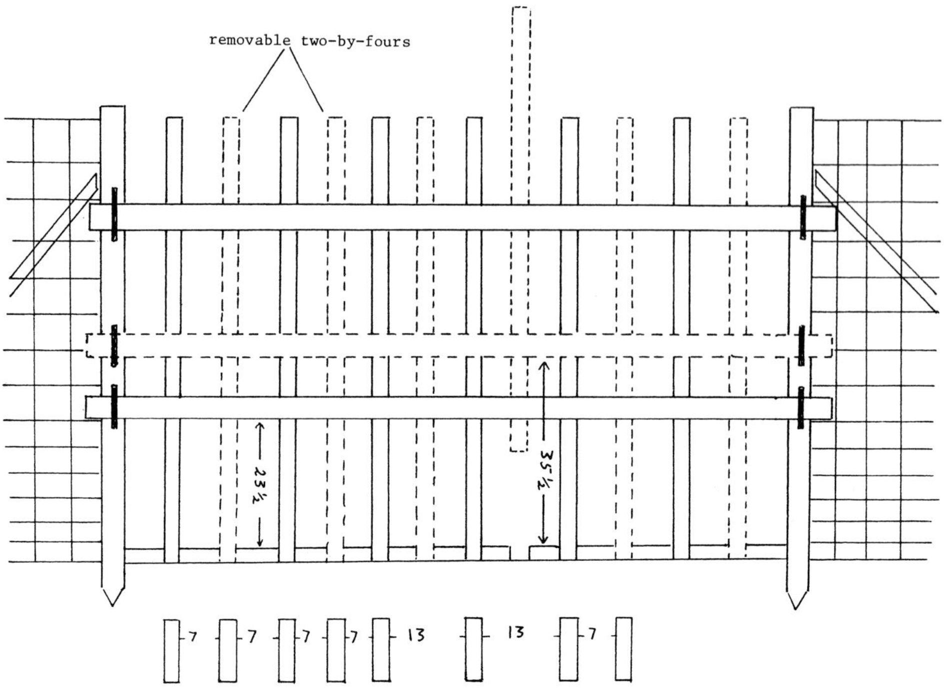

This is a fallow deer creep used at the Lucky Star Ranch for many years. The drawing is a copy of a similar drawing provided by Prof. Dr. Herman Bogner from the Bayerische Landesanstalt für Tierzucht. Until we saw this drawing, we had two different creeps at the ranch. The height of 23½ inches for the lower horizontal two-by-four has been determined by Josef Rüssmann, Germany

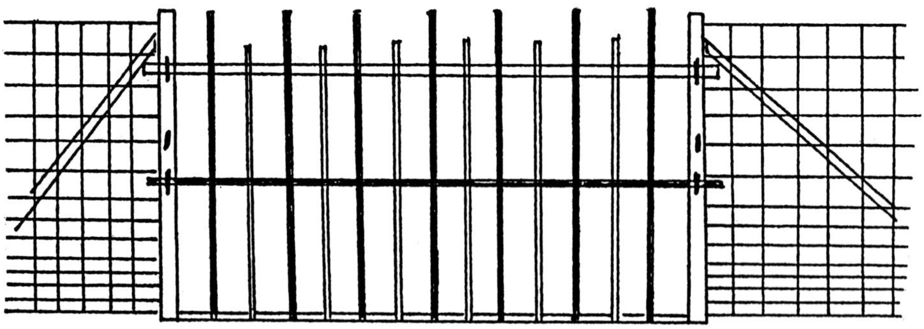

Fawn Creep

The openings of our fawn creep are 7 inches wide. The horizontal two-by- four should be 23½ inches high to keep adult Does from passing through.

Photo: Author

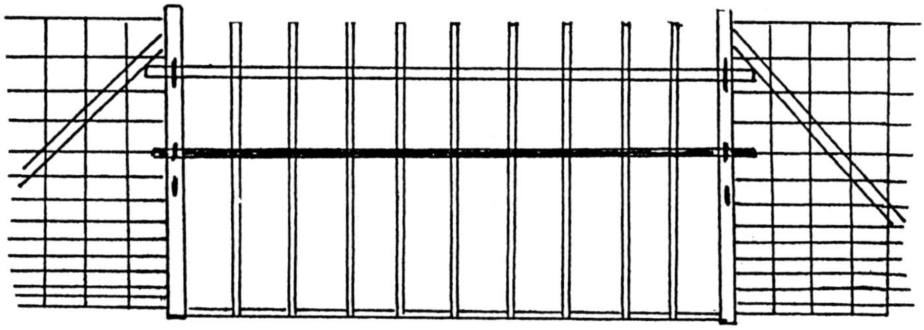

Fawn and Doe Creep

After lifting the adjustable horizontal two-by-four to 35½ inches and removing every other two-by-four to get an opening of 14 inches, Does and fawns can get through the creep while bucks are being "filtered" out.

15

Breeding

The breeding season, called the "rut," normally begins around the middle of October, sometimes a little earlier depending on latitude. The bucks, which have spent most of the summer and early fall in groups, resting and gaining strength, suddenly become rivals, even enemies. They spend a great deal of time herding their Does and marking their territories. The dominant buck performs the mating alone at first and later, a sub-dominant buck will also mate some Does, but he better watch out for the old boss. He could be in for some trouble. Fallow bucks fight intensely during mating, and will kill each other if too many bucks of the same age and stature are put together in a confined area. Deer farmers should frequently observe for injuries during the mating season, and remove wounded bucks if possible.

Proper management of bucks and Does during the mating season is one of the most important factors in successful breeding, and will result in improving the herd through genetic progress while still maintaining a high fawning percentage.

We have learned that if one pays careful attention to the principles of good mating management, and has a solid understanding of fallow deer behavior, the breeder can increase herd improvement and minimize the chance of poor fawning rates.

Bucks should be selected by frame and tameness, i.e. liveweight and temperament. Liveweight should reflect fast growth and optimum carcass weights. We have killed yearlings with carcass weights of 75 lbs. and after weighing, wished we had kept them as breeding bucks. There is definitely a connection between birth weight, weaning weight, and carcass weight. We do not have enough information yet to have a formula for this relationship.

If we could prove that the higher liveweights of the fallow deer we import from Sweden are inheritable characteristics, it would be a great step forward. In any case, we hope to have great success with the English bucks imported by Eliot Clarke, as well as with the ones we are importing from Sweden. I feel that the fresh blood lines alone will give us a great advantage in improving our breeding stock.

BREEDING RATIO

Breeding ratio is of great importance to success in breeding. Our experience shows that if one has 50 Does, for example, in one pasture during the rut, one old buck, one two-year-old buck and a spike horn would be all that is required. In case of competition, the two young bucks would stay away from the old guy, and there would be no reason to fight. Up to 40 Does can be bred by one five to seven year old buck.

In the case in which you have 100 Does in one pasture, keep four or five bucks of all different ages to keep the fighting to a minimum. Remember, unlike other deer, fallow bucks will fight to kill. When running many deer together at rutting season, we have sometimes reduced our ratio to 1:12, with two thirds older bucks and one third two-year-olds.

So far, there have been no problems reported with fertility in bucks. Don't forget to take the bucks out of the breeding herd by the middle of December to avoid late fawns.

LIVEWEIGHT AT MATING

Liveweight at mating, especially that of yearling Does at the first mating, has a large effect on reproductive efficiency. Since there is a relationship between Doe liveweight and fawn birth weight, higher Doe liveweights are important to fawn survival. This is especially true with younger Does.

The minimum liveweight of a 16-month-old Doe seems to be about 66 pounds (30 kg.) At that weight, young Does show estrus cycle and they will conceive. If the liveweight of a young Doe is above 66 pounds, the chance of survival for the fawn is more than 60%. Young Does with a liveweight below 66 pounds have only a 30% chance of raising and weaning their fawns.

We should, therefore, strive for liveweights of 75 to 85 pounds with our young breeding Does of 16 months of age. The conception rate of mature Does is normally much higher, even if their body weights are low. There is, however, a close relationship between Doe liveweight and fawn birth weight. Remember, birth weight means survival, and since fallow deer aren't that expensive to feed, there is no reason for a farmer to keep either age group under poor conditions.

Good yearling Does should, if possible, be kept in a separate pasture for mating. They should only run with an older, mature buck. The old saying "Older men make better lovers" seems to be true with fallow deer as well.

CYCLE

Sexual activity begins in males and females at approximately 16 months depending on the condition of the animals. In early October the female exhibits pubertal estrus.

At the same time the females have their first estrus of the breeding season, the bucks begin their groaning or grunting. The rutting call of fallow bucks is not a very pretty sound, not equal at all to that of the red deer stag, but the fallow Does don't seem to mind. To us it sounds more like the noise a hog would make.

The spread of first estrus occurs within a two-week period and according to Geoff Asher from New Zealand, they will reach a conception rate of 82% at their first estrus/mating. Another 13% will conceive during 2nd or 3rd estrus, and 5% may not get bred at all, if buck is removed after the second oestrus. (G. W. Asher, Ruakūra Animal Research Station, Hamilton, New Zealand). Asher also discovered that fallow deer can exhibit up to 6 estrous cycles over a five-month breeding season, and can continue to cycle until March if not successfully bred.

GESTATION

The breeding season can last from October through December, but preferably should be in October and November only. In any case, bucks should be separated from Does in early December, if possible.

The gestation period of fallow deer is approximately 234 days (+/− 4 days), and there is, according to Asher, no evidence of fawn sex, birth weight, Doe age or sire selection affecting the length of the gestation period.

CONCLUSION

Every effort should be made to cover those points that will ensure a high rate of conception and ultimately, of course, fawn survival.

1) Sires should be in good condition.
2) Doe body weight should be appropriate.
3) Vary the age of breeding bucks to reduce competition.
4) Normally, conception rates are high (92%). Keep an eye out for aborted fetuses, and if found, have them checked.

What I have told you so far in this chapter is what I consider the basic necessary information on fallow deer breeding, from my own experience. For further information, I have included the following pages of scientific work by Dr. Geoff W. Asher from Ruakara Animal Research Centre, a department of the Ministry of Agriculture and Fisheries in New Zealand. Geoff was generous enough to give us permission to reprint his work and we are very grateful to him.

The following pages have been taken from the REPRODUCTION OF FARMED FALLOW DEER (Dama dama L.)

Always remember that New Zealand has approximately the reverse seasons of North America. Just add 6 month to their timing and you will get to the right month according to our season. January 1st = June 1st, April 1st = October 1st, etcetera.

INTRODUCTION

(Excerpt from REPRODUCTION OF FARMED FALLOW DEER
by Dr. Geoff W. Asher)

Fallow deer are widespread throughout the world, being a popular game and park animal. Management of populations primarily for venison production has occurred in numerous countries including Germany (Reinken 1977), Italy (Rambotti 1985), Australia (Couchman 1980) and New Zealand.

The growth of the fallow deer farming industry in New Zealand has closely followed that of red deer farming, however, little emphasis was placed on velvet production. Initially, fallow farms were heavily concentrated around the major feral populations, particularly the northern herds at South Kaipara Heads and Wanganui. In recent years, farms have been established away from feral localities but are largely confined to the North Island. The high returns for venison and the low buying price of breeding stock relative to red deer, have been major factors behind a recent surge of interest in fallow farming.

The main factor holding back the growth of fallow deer farming is the limited availability of stock. As feral sources have been all but exhausted, more emphasis must be placed on natural recruitment within farmed populations. This paper summarises current understanding of fallow deer reproduction, an important aspect of farming the species.

(A) PHOTOPERIODIC CONTROL OF REPRODUCTION

The generally accepted hypothesis that the strict seasonality of fallow deer reproduction is mediated by the prevalent photoperiod regime (Chaplin & White 1972, Eaton 1980) has largely been inferred from experiments on red deer (Jaczewski 1954) and sika deer (Goss 1983) in which artificial photoperiod regimes were observed to alter the annual antler and testes growth cycles of males. It has been further argued that an opposite photoperiod regime is responsible for the six month displacement in reproductive seasonality of fallow deer transferred from the northern to southern hemisphere (Chapman & Chapman 1975). Experimental testing of the hypothesis has not been attempted with fallow deer.

(B) THE FALLOW DOE

PUBERTY

The general age of fallow Does at first mating has been variously assessed as 16 months (Harrison & Hyett 1954; Ueckermann & Hansen 1968; Armstrong *et. al.* 1969; Chapman & Chapman 1969, 1975; Baker 1973; Reinken 1977; Asher *et. al.* 1981; Asher & Adam 1985). Occasional well-grown individuals have conceived at 4–5 months of age (personal observation).

The threshold liveweight for attainment of puberty has been difficult to assess. Data on lactational status of 27 month old Does (first weaning) on several NZ farms indicated that most Does below 30 kg at 16 months of age failed to subsequently rear a fawn. However, direct observations of mating suggest that puberty weight, in terms of first oestrus, is probably around 28 kg. As average liveweights in NZ are between 34–38 kg at 16 months of age, most Does are likely to be well above the critical puberty weight.

Chapman & Chapman (1975) and Reinken (1977) considered that pregnancy and fawning rates are usually lower for pubertal Does than for older Does. However, data from NZ farms indicate rates comparable to those of older Does (Asher 1985). First fawning Does (24 months of age) generally fawn about a week later than older Does, presumably reflecting a slight lag in first oestrus at puberty.

ONSET AND SYNCHRONY OF OESTRUS

The onset of the fallow deer breeding season is well defined in terms of first overt oestrus, which appears to correspond exactly with the onset of male rutting vocalisations. There is also little variation between years; on Ruakura the first oestrus Doe can be expected between 15–20 April each year. (USA—15–20 October)

Oestrus is well synchronised within a herd. The spread of first oestrus on Ruakura in 1983 and 1984 was only twelve days, with pubertal Does being later on heat than older Does.

SILENT OVULATIONS PRECEDING FIRST OESTRUS

Progesterone profiles from daily blood sampling have indicated the occurrence of silent ovulations resulting in transient corpora lutea before first overt oestrus of the breeding season. This was confirmed by laparoscopic examination of Does before first oestrus. Some Does appeared to have two or more such ovulations in succession (Figure 1,a). The occurrence of short-lived corpora lutea (10–12 day cycles) during the transition of the breeding season helped to synchronise first oestrus, as the spread of first oestrus corresponded to the longevity of the short cycles.

LENGTH OF THE OESTROUS CYCLE

Although originaly reported as being monoestrous (Harrison & Hyett 1954; Hamilton, Harrison & Young 1960; Asdell 1964), fallow Does are now known to be seasonally polyoestrous (Cowan 1965; Armstrong *et. al.* 1969; Chapman & Chapman 1975). Cowan (1965) claimed the oestrous cycle length was 24–26 days for Does in British Columbia and Baker (1973) estimated it to be 26 days in NZ. Both observations were based on only one or two individuals.

A total of 34 Does were run continuously with harnessed (Sire-sine) vasectomised bucks on Ruakura. Oestrous was detected on 177 occasions, resulting in a total of 142 valid oestrous cycle length observations. Mean

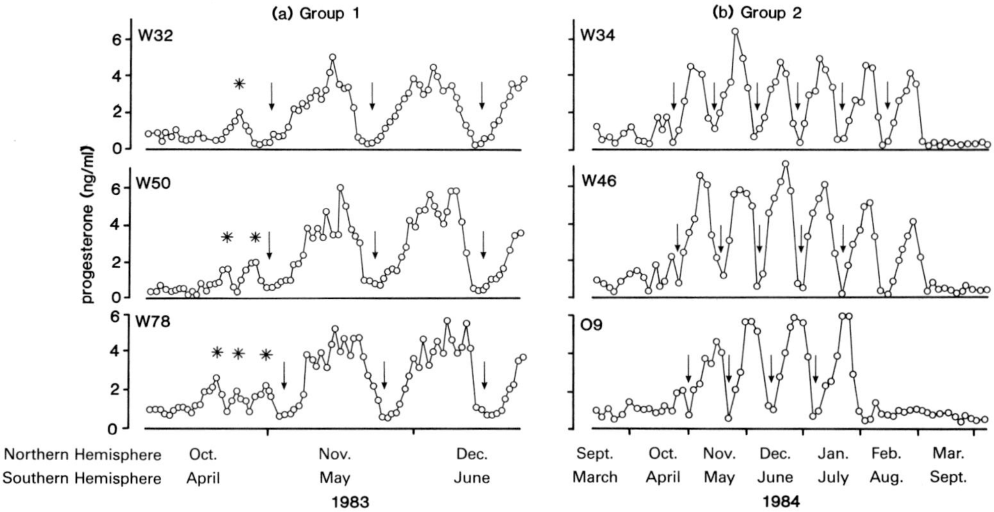

Figure 1
Selected typical profiles of serum progesterone levels for; (a) three Group 1 fallow Does showing evidence for the occurrence of single or multiple short-lived corpora lutea as indicated by transitory elevations of serum progesterone levels before first oestrus, and (b) three Group 2 Does showing different events terminating the breeding season. Arrows indicate day of oestrus.

oestrous cycle length was 22.4 days ± SD 1.3 with a range of 20 to 27 days. Cycle length increased as the breeding season progressed, with the differences between the means for the first, second and third cycles being significant ($P < 0.05$), thereafter, they became more variable (Table 1). Year, Doe age and Doe liveweight did not significantly ($P > 0.1$) affect oestrous cycle length (by cycle number).

Table 1
Fallow deer oestrous cycle length by cycle number

Cycle no.	Observations	Cycle length (days)	Standard deviation (days)
1	33	21.0	0.64
2	33	22.0	0.66
3	33	22.9	0.97
4	28	23.0	1.11
5	12	23.5	1.45
6	3	25.7	1.53

A noteable feature is the low variance encountered for the length of the first three oestrous cycles (Table 1) compared with respective data for red deer (Guinness, Lincoln & Short 1971).

DURATION OF THE POTENTIAL BREEDING SEASON

In the absence of conception, cyclic activity of Does continues throughout winter and into early spring. This is well illustrated by progesterone profiles of Does run with a vasectomised buck (Figure 1,b). The number of observed oestrous cycles varied from 3–6 per Doe, with the duration of the oestrous-bound breeding season varying from 65–135 days. Pubertal Does had fewer cycles than older Does and Does in their second breeding season had fewer cycles than older Does (Table 2).

Table 2
Duration of the potential breeding season (number of oestrous-bound cycles) by Doe age.

Doe age (months)	No. of Does	Mean no. of cycles (SD)	Mean date of first oestrous (SD days)		Mean date of last oestrous (SD days)	
16	9	3.56 (0.53)	2 May (2.8)	Nov.	20 July (12.4)	Jan.
28	17	4.24 (0.56)	27 April (4.5)	Oct.	30 July (12.5)	Jan.
40+	7	5.43 (0.54)	24 April (3.3)	Oct.	25 Aug. (15.1)	Feb.

The age effect was primarily due to the variation in occurrence of last recorded oestrous (Table 2). All Does cycled beyond the shortest day, but younger Does may have a more sensitive response pattern to increasing day length than older Does, resulting in earlier anoestrus.

CONCEPTION RATE AT FIRST OESTRUS AND RETURNS-TO-SERVICE

The conception rate at first oestrous appears to be high for fallow Does on the basis of the synchrony of fawning (Asher & Adam 1985). The conception rates and returns-to-service for fallow Does observated mated on Ruakura are presented in Table 3.

Table 3
Summary of conception rates at consecutive oestrous periods.

	Observed 1st oestrous matings	Minimum conceptions at 1st oestrus	Minimum conceptions at 2nd oestrus	Minimum conceptions at 3rd oestrus	Non-pregnant Does
1983	42	33 (78.6%)	6 (14.3%)	2 (4.8%)	1 (2.4%)
1984	56	48 (85.7%)	5 (8.9%)	(Buck removed early)	3 (5.4%)
TOTAL	98	81 (82.7%)		13 (13.3%)	4 (4.1%)

The above data tend to support the fawning data, suggesting a high initial conception rate at first mating. Removal of bucks before third oestrous is likely to result in a slightly higher non-pregnancy rate, however, will prevent fawns being born after mid-January. (USA—mid-July)

GESTATION LENGTH

The gestation length of fallow Does has been estimated at 229 days (Chapman & Chapman 1975). The gestation length of Ruakura Does in 1983 and 1984, based on 88 observations, was 234.2 days (SD 2.7 days). There was no effect of Doe age, fawn sex, fawn birth weight, year and sire upon gestation length ($P > 0.1$, ANOVA)

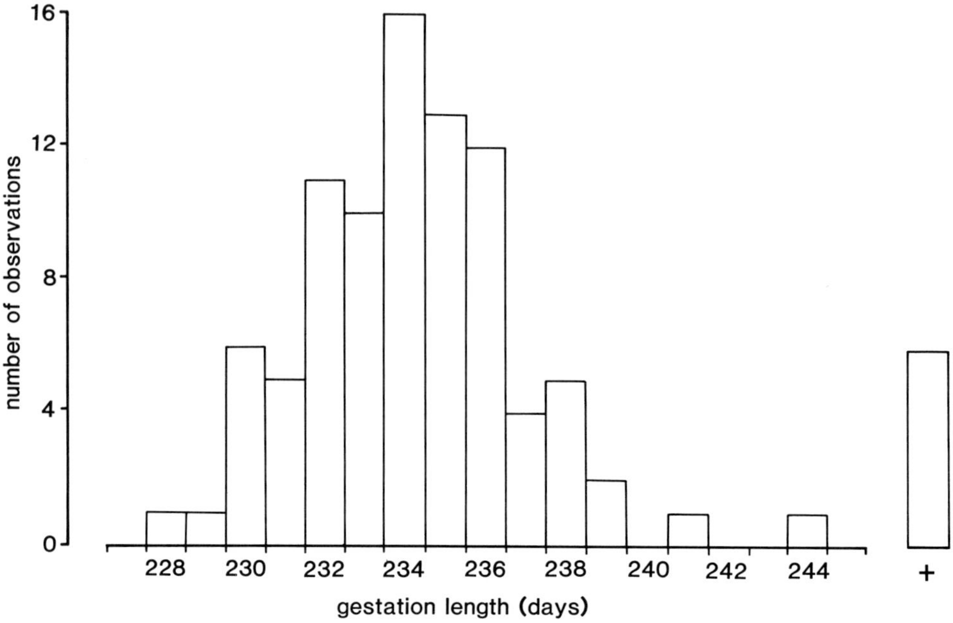

Figure 2

Histogram showing frequency distribution of 94 observed gestation lengths (days) of fallow Does in 1983 and 1984. Six observations were beyond one oestrous cycle length (21 days) from the modal gestation length (234 days) and are indicated at the right hand side of the histogram.

PROGESTERONE PROFILE OF PREGNANT AND NON-PREGNANT DOES

The progesterone profiles of 55 pregnant and 14 non-pregnant Ruakura Does in 1983 are presented in Figure 3.

During the season of cyclic activity of the non-pregnant Does, mean progesterone levels fluctuated, clearly showing individual oestrous cycles even though the data are not pooled by day of oestrous. Only at the cessation of all cyclic activity in late spring was there a significant differentiation of progesterone levels between pregnant and non-pregnant Does. For valid pregnancy diagnosis, determination of progesterone levels in single samples will only be useful between mid-October (post-cycling) (USA—April) and mid-November (USA—May) (pre-fawning). Prior to this, extremely low progesterone levels (<0.5 ng/ml) may well indicate non-pregnancy, but high levels (>4.0 ng/ml) do not necessarily indicate pregnancy.

FAWNING SEASON

The fallow deer fawning season on four NZ farms (1980/81–1983/84 seasons) is illustrated in Figure 4. Birth dates were obtained by tagging all fawns within 12 hours of birth. The data are pooled between farms and years

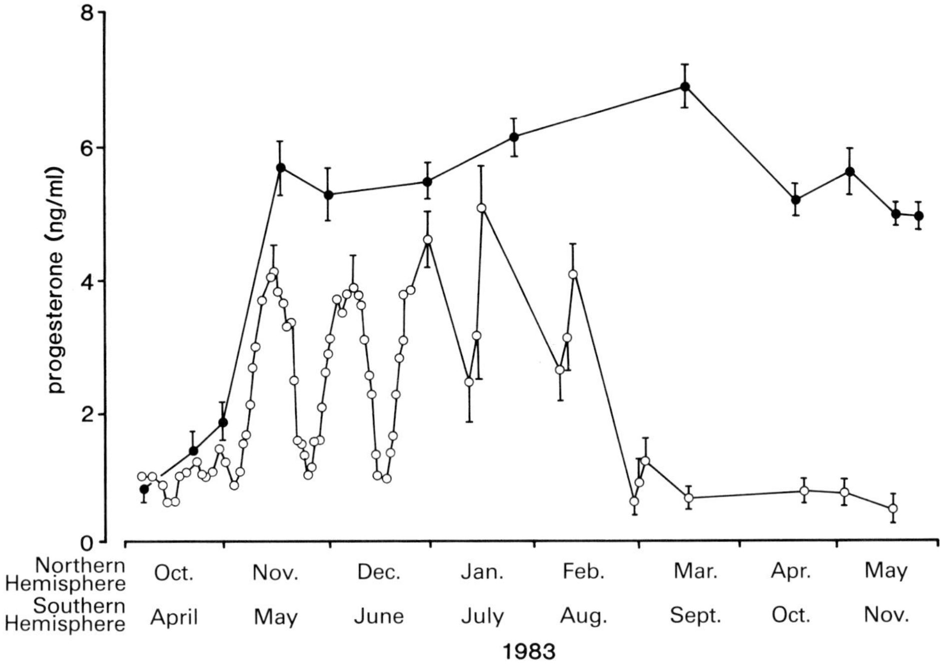

Figure 3

Profiles of mean (±sem) serum progesterone levels, between April and November 1983, for 55 pregnant (●———●) and 14 non-pregnant (○———○) fallow Does.

without adjustment for variation in the onset of fawning and median fawning date.

The fawning season was more synchronised than that of red deer (Asher & Adam 1985) reflecting both a synchronised first oestrous and high conception rate at first oestrous. The distribution of births was, however, skewed to the right and it is probable that the small proportion of later born fawns were a result of return-to-service conceptions. Occasional fawns were born as late as March (USA—Sept.) and were, therefore, conceived in July (USA–Jan.) (fifth or sixth oestrous).

REPRODUCTIVE RATES

On-farm reproductive rates were obtained from a number of Waikato/Bay of Plenty fallow farms for the 1980/81–1983/84 fawning seasons. The data were obtained by postal questionnaire and from farms monitored daily during fawning (Tables 4 and 5).

Overall reproductive rates (Table 4) were not as consistent between years as for red deer. Estimated fawning and weaning rates were lower, and fawn mortality rates higher than for red deer (Asher & Adam; MAF, Agricultural Research Division annual report, 1983/84).

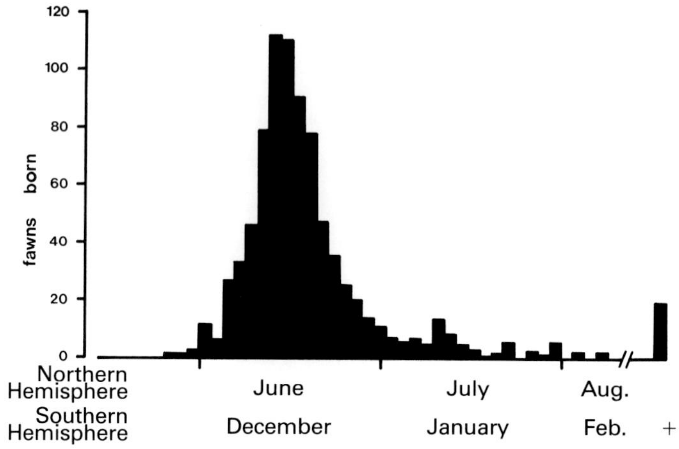

Figure 4

Histogram showing the distribution of fallow deer birth dates recorded on four Waikato/Bay of Plenty farms during the four fawning seasons from 1980/81–1983/84 (after: Asher, 1984; Asher, 1985b).

Table 4
Reproductive performance of fallow Does in some northern NZ regions.

Season	Farms (no.)	Does (no.)	Fawning Rate %	Weaning Rate %	Fawn Mort. %	Abortion Rate %	Assisted Fawning
1980/81	25	519	82.7	69.6	15.9	—	—
1981/82	29	525	84.2	72.8	13.6	4.2	1.1
1982/83	22	551	75.9	65.7	13.4	3.6	0.0
1983/84	27	752	80.9	69.5	14.0	0.7	0.3

Table 5
Reproductive performance of fallow Does on monitored and survey farms.

	1981/82		1982/83		1983/84	
	Monitored	Survey	Monitored	Survey	Monitored	Survey
Farms (no.)	3	26	3	19	3	24
Does (no.)	210	315	222	329	236	516
Calving rate (%)	91.4	79.4	90.5	65.4	92.8	75.4
Weaning rate (%)	77.6	69.5	74.3	59.3	75.0	67.1
Fawn Mort. (%)	15.1	12.4	17.9	9.3	19.2	11.1
Abortion rate (%)	5.7	3.2	4.5	3.0	0.9	0.5
Minimum conception rate (%)	92.8	—	95.0	—	93.7	—

Monitored farms collectively contained a large proportion of the total Doe population sampled (Table 5) and also had higher reproductive rates, despite higher fawn mortality, than survey farms. It is likely that, without close supervision of herds during fawning, farmers on the survey farms underestimated both fawn losses (and hence fawning rates) and abortions.

Combining fawning rates and abortion rates from monitored farms indicated that the herd minimum conception rates were high (Table 5), clearly showing that Doe infertility (barreness) is unlikely to be a major cause of reproductive failure on most farms. This was further supported by data from recent slaughtering of 70 fallow Does from a herd of 3000 Does. These particular animals were assessed as non-lactating in March (USA—Sept.) (weaning) for three years. Upon slaughter in June/July (USA—Dec.–Jan.) 1984, all but ten (0.3% of the total herd) were pregnant. The ten non-pregnant animals included five cycling Does with encysted foetuses and five non-cycling Does with vestigial reproductive tracts.

Clearly, more detailed investigations of fawn mortality and abortions are warranted.

FAWN MORTALITY

The causes of fawn deaths on the monitored farms from 1980–84, as determined by post-mortem examination (after McFarlane 1965) are summarised in Table 6.

Table 6
Fallow fawn mortality (1980–84).

Cause of death	Number of deaths	%
Non-viability	40	24.9
Starvation	31	19.3
Dystocia	23	14.3
Misadvanture	18	11.2
Gut infection	16	9.9
Throat/jaw infection	11	6.8
Lung infection	6	3.7
In utero death	5	3.1
Liver infection	3	1.9
Hypothermia	2	1.2
Abnormality	1	0.6
Unexplained	5	3.1

Non-viability (undersized fawns that failed to walk) accounted for a quarter of recorded deaths. This was further reflected in the relationship between birth weight and fawn mortality (Table 7).

Table 7
The influence of birth weight on fawn mortality.

Birth weight (kg)	<3.0	3.0–3.9	4.0–4.9	5.0+
No. of fawns	94	438	281	9
No. of deaths	56	71	30	1
Mortality %	59.6	16.2	10.7	11.1

Further investigation of the problem has shown that first fawning Does (2 year olds) tended to produce lighter fawns at birth, *relative to their own liveweight*, than older Does (Asher & Adam 1985). Mortality of fawns from first fawners was as high as 35%.

Fawn deaths due to starvation and fawning difficulties (dystocia), although accounting for 19% and 14% of fawn deaths respectively, were considerably lower than for red deer.

Death of fawns through misadventure (usually fence hangings) have been disappointingly high on some farms but were mostly overcome by techniques of fawn proofing paddocks.

Occasional outbreaks of infectious agents (enteritis, jaw infections) resulted in higher fawn mortalities on some farms in the 1983/84 season. In extreme cases of enteritis, new born fawns were effectively treated with broad-spectrum antibiotics prior to showing symptoms. This necessitated handling at birth, a practice not common on deer farms.

ABORTIONS

There is some circumstantial evidence that third trimester abortions on some fallow farms were due to recurrent leptospirosis infection (*hardjo* & *pomona*). Several farmers appear to have overcome the problem by treating all stock with antibiotics (Streptopen) and a biannual vaccination (Leptovoid).

The abortions do not appear to relate to yarding activity, as has been suggested on many occasions.

(C) THE FALLOW BUCK

PUBERTY

Puberty in fallow bucks has been defined as the period between the commencement of male hormone production and the formation of spermatozoa by the testes (Chapman & Chapman 1975).

Two principal androgens; testosterone and androstenedione; were measured from serum samples taken from entire and castrate bucks at 2–3 week intervals. The profiles (Figure 5) show that in the entires, androgens increased in concentration from January (USA—July) (12–13 months of age), reached peak mean levels in March (USA—Sept.) (15 months) before the rut and declined appreciably at the start of the rut (mid April) (USA—mid-October). Moderate levels were maintained through winter.

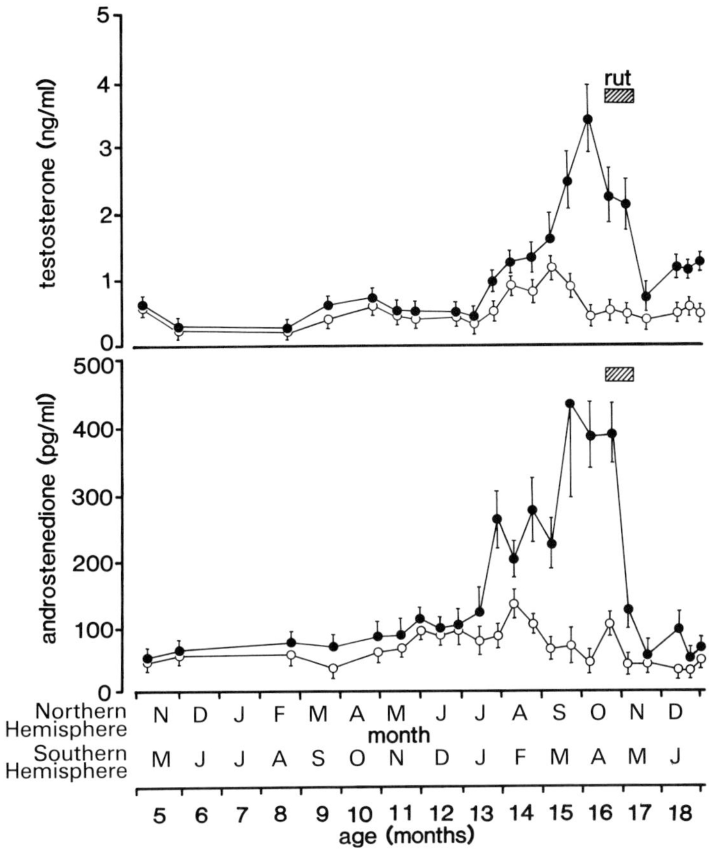

Figure 5

Profiles of mean (±sem) serum testosterone and androstenedione levels, from 5 to 18 months of age, for 21 entire (●———●) and 11 castrated (○———○) fallow bucks.

The castrates showed a small peak in both testosterone and androstenedione coincidental with that of the entires (Figure 5). As this peak did not occur in subsequent years it is probable that, not only were the androgens of extra-gonadal origin, but were linked with the onset of puberty, and may normally supplement gonadal androgens.

The morphology of the testes of variously aged bucks has been described in relation to puberty (Chapman & Chapman 1969, 1970, 1975, 1980; Chaplin & White 1970, 1972; Baker 1973). The accounts closely agree that primary spermatocytes became abundant between 6 and 10 months of age, secondary spermatocytes were evident at 11–12 months and all stages of spermatogenesis present at 13–14 months. In addition, Chapman & Chapman (1980) described the morphological development of the male accessory organ, and clearly showed that full functional development occurred by 16 months of age.

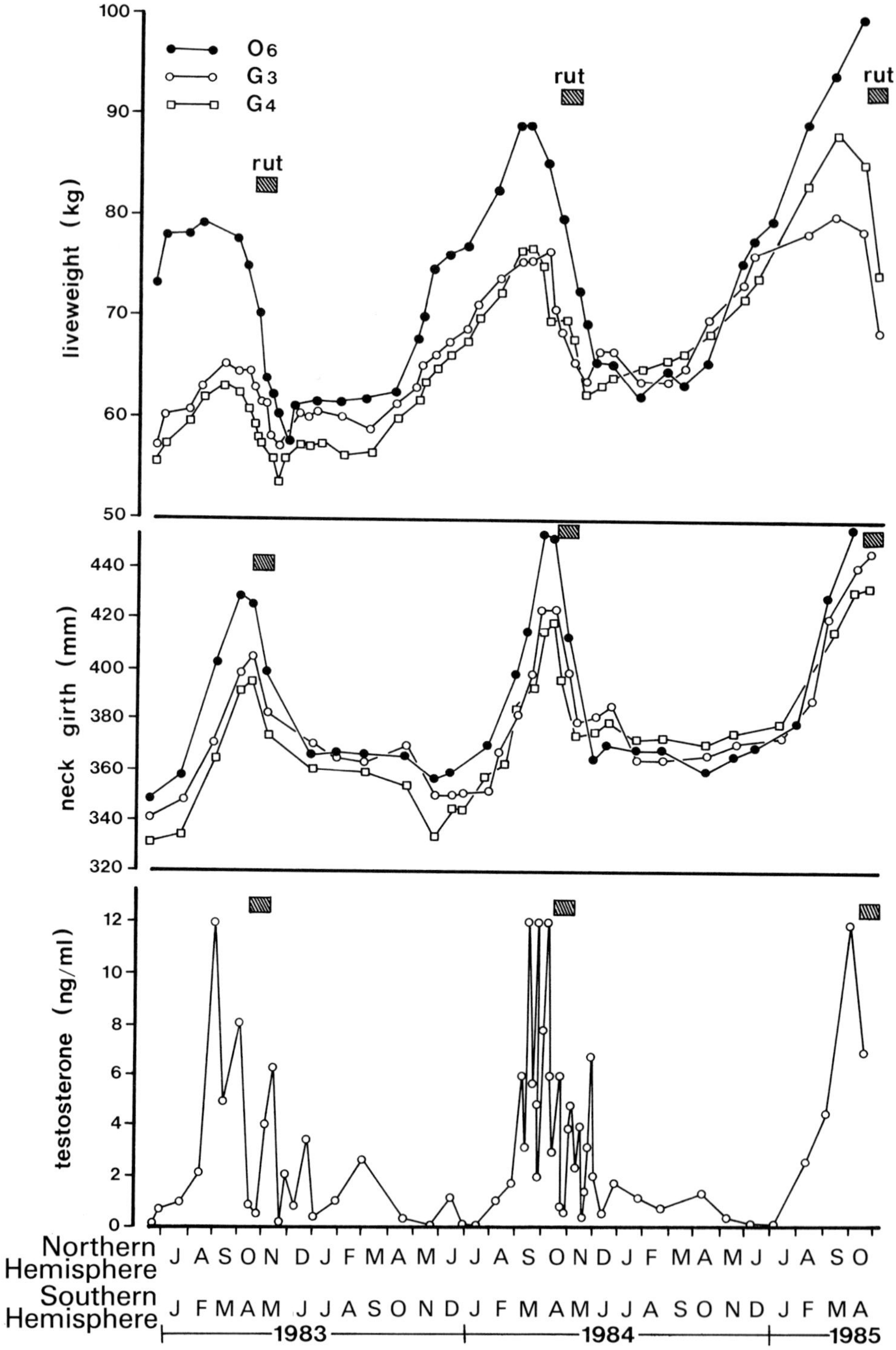

Figure 6

Seasonal profiles of liveweight and neck girth of three sire fallow bucks (Or6, G3, G4), recorded at fortnightly or monthly intervals from December 1982 until April 1985 (29 months) and an example of the seasonal profile of serum testosterone levels of one of these bucks (G3).

REPRODUCTIVE SEASONALITY OF MATURE BUCKS

The dramatic annual liveweight cycle of mature male deer is well known. It is particularly pronounced in fallow bucks used as herd sires (Figure 6).

In the example above, liveweight showed a steady increase from October (USA—April) and peaked in March (USA—Sept.). Over eight weeks spanning the period of intensive rutting actively (late April/early May) (USA—Oct.–Nov.), individual bucks lost up to 25% of their March (USA—Sept.) weight. Although there was some post-rut recovery the bucks continued through winter and early spring (June–October) (USA—Dec.–April) at low liveweights and, presumably, with depleted fat reserves. Liveweight recovery did not start until October (USA—April), about a month later than for red deer stags (Ruakura data).

The reproductive seasonality of a group of 21 polled (early pedicle cauterisation), entire bucks, run as a bachelor herd, was continually monitored from puberty (16 months) through to 40 months of age. A twelve month profile (22–34 months) of liveweight, testosterone, neck girth, testis diameter, sperm density and sperm motility was compiled for seven of these bucks (Figure 7). The semen was collected by electro-ejaculation of fully anaesthetised individuals.

The annual cyclical pattern of liveweight was not as pronounced as for the herd sires and may be due to the absence of oestrous Does. Testosterone levels were basal for the months of November and December (USA—May & June) and correlate with complete aspermatogenesis and testicular/neck regression. With antlered bucks, this period is characterised by antler casting, followed by rapid growth of velvet antler. Increasing testosterone levels through January–March (USA—July–Sept.) (late summer–early autumn) were associated with progressive spermatogenesis (as reflected by sperm density and motility), testicular enlargement and neck expansion. Rapid calcification and stripping of antlers would normally occur in February (USA—Aug.).

Testicular activity (sperm production, testosterone secretion) continued throughout the period from April–October (USA—Oct.-Apr.). Curiously, sperm density and motility were maximal *after* the rut (April/May) (USA—Oct.–Nov.), whereas, neck girth and testis diameter did not increase beyond the rut. This may represent early completion of maturation of secondary sexual characteristics in response to very high pre-rut testosterone levels, which may be partially inhibitory to maximal sperm production early in the breeding season. Similarly, in a recent trial a buck was observed to show poor mounting responses to Does induced into oestrous (progesterone CIDR and PMSG) only three weeks before the onset of the natural rut. This may have important implications with respect to artificial manipulation of the breeding season of fallow deer.

Conversely, during studies on oestrous cycle length of fallow Does, vasectomised bucks were observed mating oestrous Does as late as October (USA—Apr.), although there is some suggestion that libido may have been declining at this late stage.

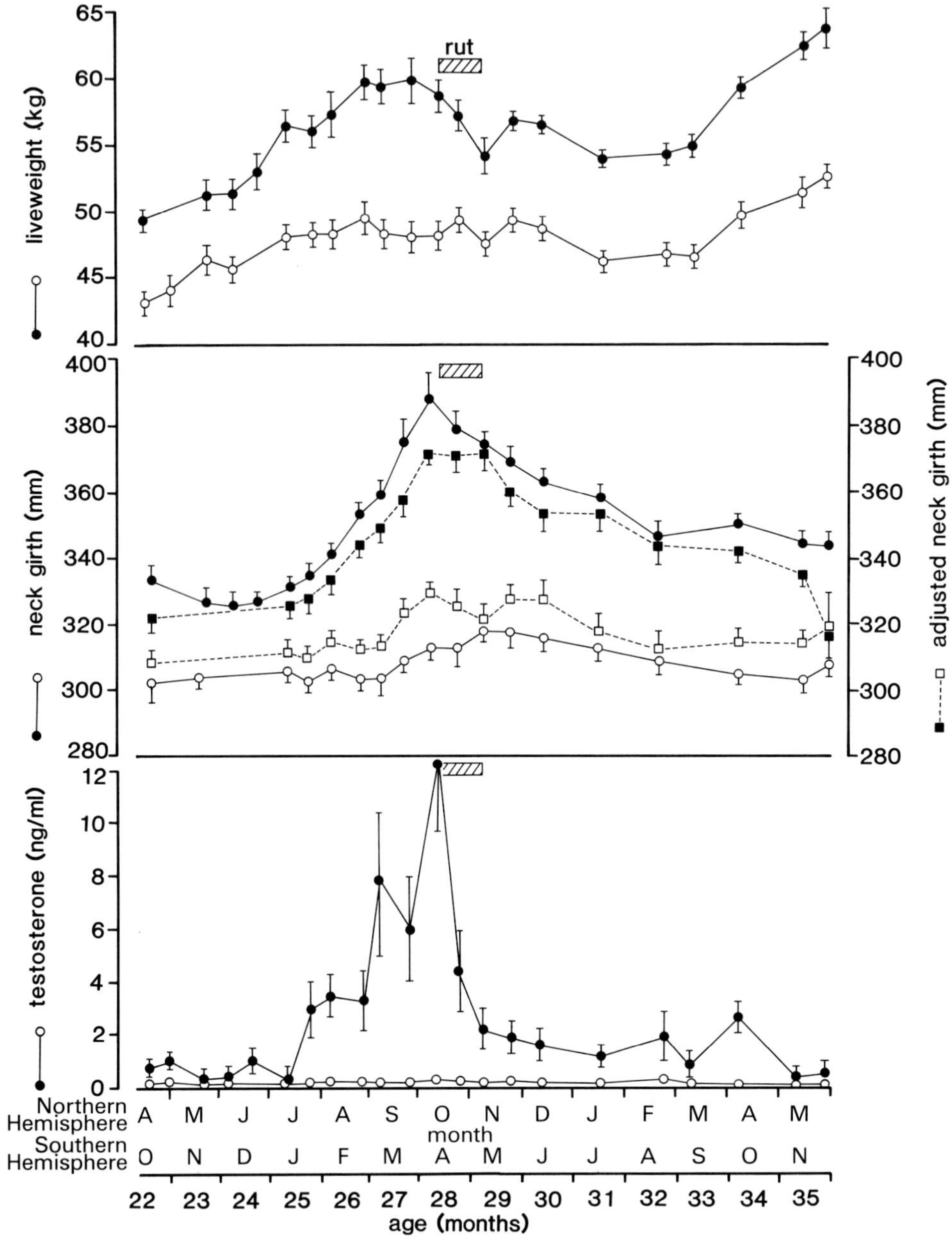

Figure 7

Seasonal profiles of mean (±sem) liveweight, non-adjusted or adjusted neck girth and serum testosterone levels, from 22 to 35 months of age, for 7 entire (solid symbols) and 11 prepubertally castrated (open symbols) fallow bucks.

BUCK ACTIVITY DURING THE RUT

The term "rut" applies to the period of intense overt reproductive activity of the buck. The rapid loss of liveweight observed in rutting fallow bucks (Figure 5) is associated with major short-term changes in daily activity during this period.

Regular observations of activity were made on three sire bucks (as for Figure 5) in 1983. Spot observations of major activities (grazing, sitting, walking, running) were conducted every 15–30 minutes over 24 hour periods. Groaning (male vocalisations) was recorded additional to these major catagories as it occurred throughout all activities (even grazing!).

Figure 7 shows changes in both grazing and walking activity through the rut. The frequency of total observations in which groaning was observed, and the frequency of observed oestrous/matings, are included.

The three bucks showed similar changes in their pre-rut, rut and post-rut activities. Clearly, groaning activity was closely associated with the presence of oestrous Does. Grazing activity decreased from approximately 40% of pre-rut observations to as low as 5% by the start of the groaning/mating period. Correspondingly, the bucks became considerably more active, in terms of mobile activities (walking, running), as the frequency of oestrous Does increased. This was due to considerable effort in the patrolling of territorial boundaries (usually fence-lines) and frequent chasing of Does.

As the frequency of oestrous Does decreased in early May (USA—Nov.), the groaning ceased and mobile activities returned to normal levels. However, there was a pronounced increase, compared to pre-rut levels, of grazing activity, presumably in response to intense hunger and severe depletion of energy reserves.

In all, the major rutting period lasted for only two weeks but had a profound effect upon the bucks' liveweight (Figure 6).

MATING BEHAVIOR

During the 1983 mating season 45 matings were observed, of which 21 (46.7%) were mating sequences recorded during daylight hours. The remaining 24 (53.3%) matings were observed, but not formally recorded, during night-time. The main behavioural characteristics of oestrus (female component), courtship (male component) and copulation are described as follows:

Pre-oestrous behaviour of Does and responses by the buck. Although not formally recorded prior to mating sequences, the first indication of impending oestrus was often observed approximately 12 to 24 hours before its onset, when the Doe disassociated itself from herd-mates, continually paced fence-lines and frequently exhibited aggression towards any encroaching Does. During this period, the buck periodically approached the Doe, often displaying an exaggerated "goose-step" walk, flared pre-orbital glands and raised tail. However, the Doe actively avoided the buck during such encounters

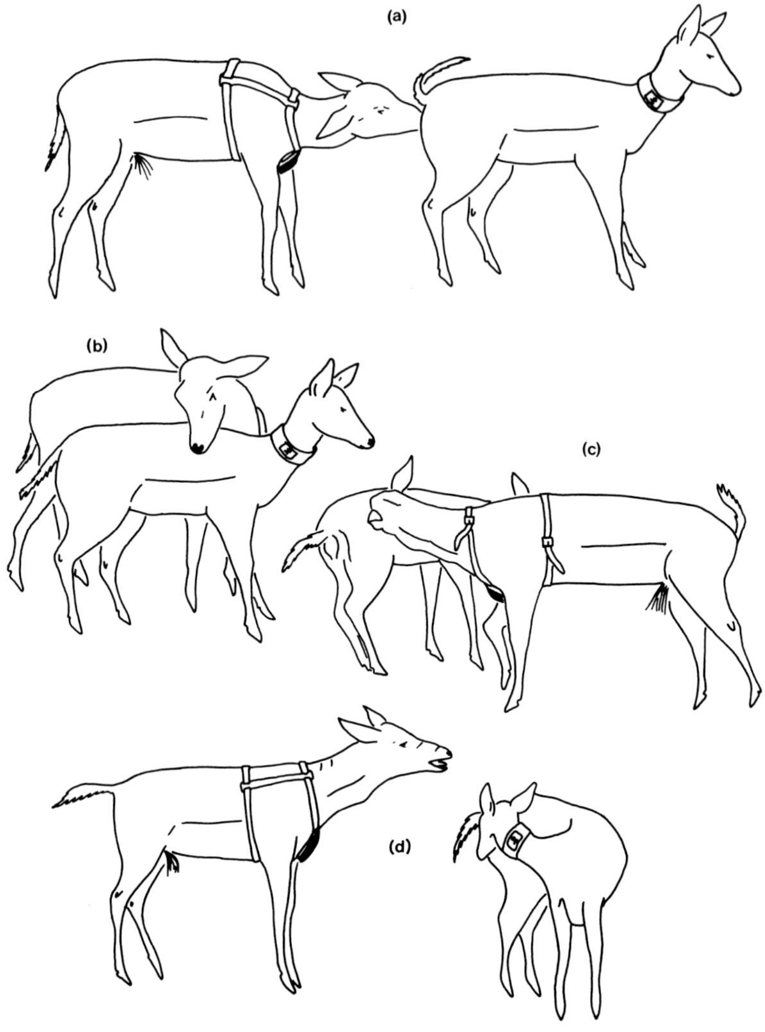

Figure 8

Behavioural events observed during mating sequences of fallow deer.

(a) buck sniffing vulval region of an oestrous Doe
(b) chin over shoulder stance exhibited by the buck
(c) buck bunting an oestrous Doe following a previous confrontation with a rival buck
(d) buck groaning while the oestrous doe exhibits characteristic preening activity

and was often chased back into the main body of the herd. Characteristically, the Doe would run with its neck and head inclined below the level of the shoulders and would bleat repeatedly until the chase was over. The buck usually emitted several loud groans during the chase and, once stationary,

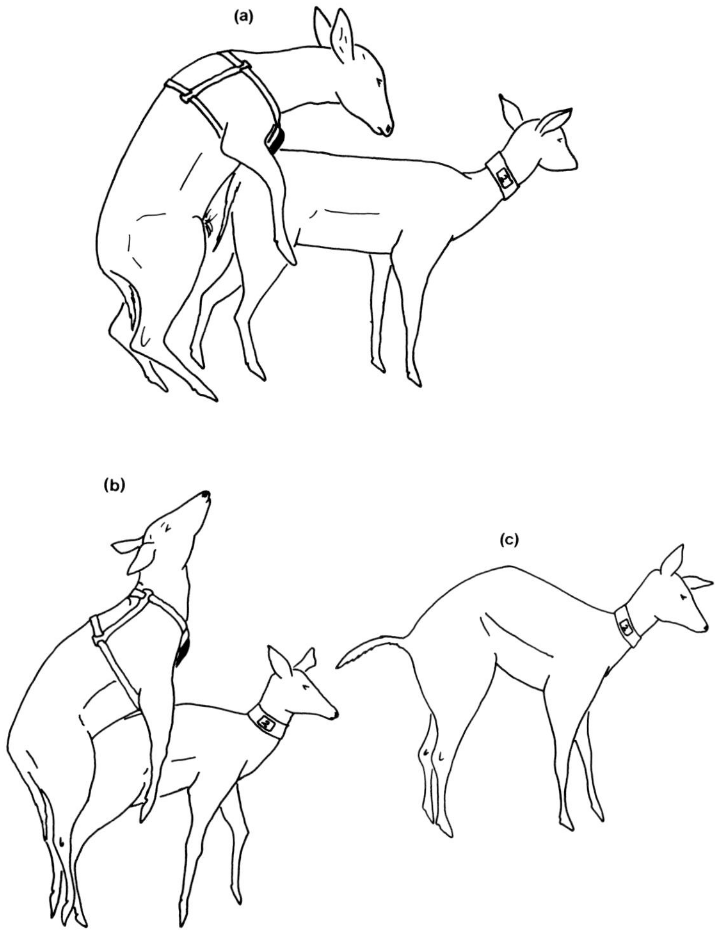

Figure 9

Behavioural events observed during mating sequences of fallow deer.

(a) non-copulatory mount
(b) copulatory thrust following intromission
(c) post-copulatory stance exhibited by the Doe

would often continue groaning. Such sequences were suggestive of herding behaviour, as described for fallow deer by Cook (1978).

Pre-oestrous behaviour of Does continued to within a few hours of the overt oestrus. However, during the intervening period the Doe spent much of the time resting (sitting). During this stage, the buck would often remain close to the Doe, sometimes even resting just a few metres away. Usually, however, the buck made frequent olfactory checks of the resting Doe's vulval region during routine investigatory walks. In response to this, the Doe either remained sitting, exhibited a low neck stretch parallel to the ground and

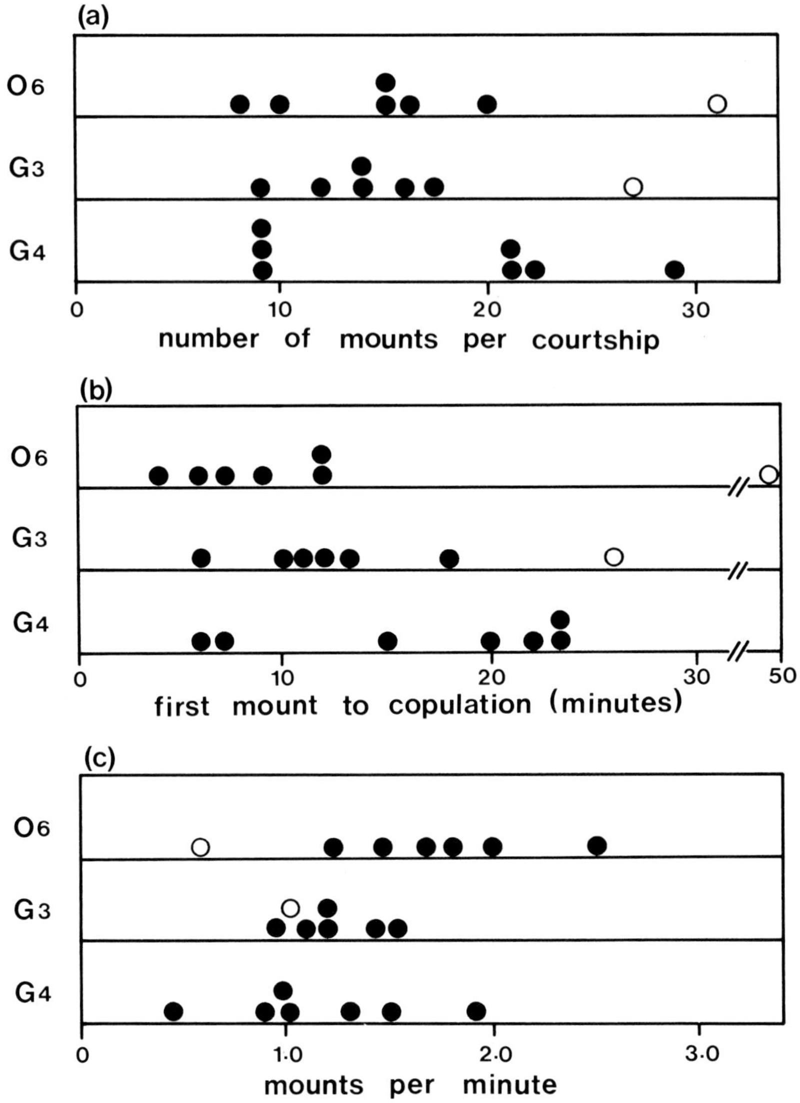

Figure 10

Frequency plots of (a) the number of mounts per courtship, (b) the interval from first mount to copulation and (c) the mount rate per courtship, for seven recorded mating sequences for each of the three sire bucks in 1983. The open circles represent observations (n = 2) in which there was repeated disruption of mating sequences by rival bucks.

either bleated repeatedly or stood up and ran away, only to sit down again in another part of the paddock.

Onset of oestrus and courtship. The first, usually somewhat abrupt, indication of overt oestrus was either the initiation of fervent self-grooming (preening) activity by the standing Doe (i.e. continual chin-rubbing over the

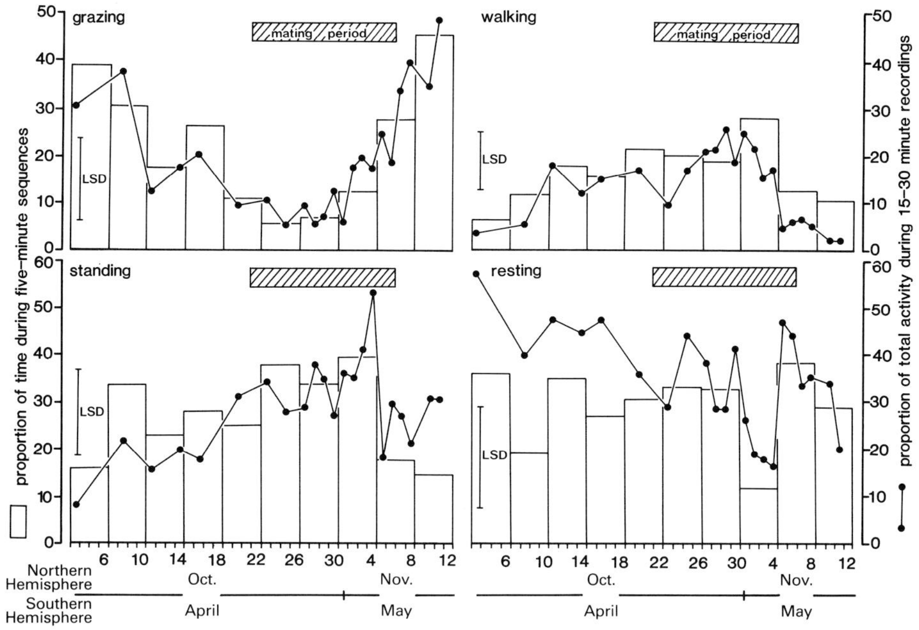

Figure 11

Variation, throughout the mating season, in major primary activity allocations (i.e. grazing, walking, standing, resting) of three fallow bucks during the period spanning matings, as indicated by; (i) the proportion of total primary activity during five-minute activity sequences for four-day groupings (bar graph) (LSD = least significant difference); and (ii) the proportion of total primary activity during 15–30 minute interval observations over 24 hour periods (●———●). All data are the pooled value for all bucks and diurnal periods.

rump and back; Figure 8d) or a sudden approach by the buck that was not associated with submissive/avoidance behavior by the Doe. In the latter case, the buck generally showed considerable interest in the Doe by sniffing or licking the Doe's vulva (Figure 8a), assuming a characteristic stance with its chin over the Doe's shoulder or back (Figure 8b), grooming to Doe's face and ears, or groaning nearby (Figure 8d). In response, the Doe usually remained standing (Figure 8a and 8b) or continued preening (Figure 8d). Such activity usually persisted for only a few minutes before the first tentative mount was made by the buck (Figure 9a). As the transition to overt oestrus and mounting activity was so abrupt, the first mount was occasionally the first indication of oestrus that was noted by observers.

Courtship. Repeated mounting of the oestrous Doe was a common feature of all recorded mating sequences. In most cases only the final mount was a copulatory mount (Figure 9a). Between 8 and 31 mounts were recorded during courtships (Figure 10a) and the interval between the first tentative mount and the final copulatory mount (Figure 10b) ranged from 4 to 50 minutes. The mount rate (Figure 10c) varied from 0.6 to 2.5 mounts per minute. There were no significant differences between the three sire bucks for either the number of mounts per courtship, the courtship interval or the mount rate per courtship.

The early mounts by the buck during courtship sequences were generally not associated with protusion of the penis from the preputial sheath (Figure 9a). However, as courtship proceeded, the penis became progressively more protruded and erect. Throughout courtship the Doe continued to exhibit characteristic displays of oestrus, including intense preening activity and, very occasionally, grooming of the buck's face, neck and shoulder.

During courtship neighbouring bucks sometimes showed considerable interest from the other side of the boundary fence. This involved considerable fence-pacing and groaning activity. On these occasions, the courtship buck was frequently distracted from mating and fights occurred between the two rivals (i.e. head to head clashes through the boundary fence). After each bout of fighting, the courting buck returned to the oestrous Doe and resumed courtship, sometimes exhibiting aggression toward the Doe (e.g. bunting the Doe's flanks; Figure 8c) followed by mounting. Interference of courtship by a rival buck occurred twice during recorded mating sequences, apparently resulting in a prolonged courtship interval, an increased number of mounts and a generally lower mount rate than for other mating sequences (Figure 10).

Copulation. The final copulatory mount (Figure 9b) was distinctly different from all other mounts. Following intromission the buck gave a violent pelvic thrust, with its hind legs leaving the ground (as has been frequently described for red deer; Guinness, Lincoln & Short, 1971; Clutton-Brock, Guinness & Albon, 1982; Veltman, 1985). The Doe was generally propelled forward by the force of the thrust and, on two occasions, was observed to collapse momentarily beneath the buck.

Post-copulatory behaviour. Following copulation the buck and Doe moved apart. The Doe usually disassociated itself from herd-mates and stood motionless for between 5 and 120 minutes with its back arched and tail raised (Figure 9c), occasionally straining abdominal muscles and passing clear mucus from the vulva. Some Does frequently licked their vulva region during the post-copulatory period. The buck usually urinated immediately following copulation, some groomed its penile region and seldom showed further interest in the Doe.

On only two occasions out of 45 observed matings (i.e. 4.4%) were Does observed to continue exhibiting oestrus following copulation. In both cases copulation occurred twice within six hours. Therefore, behavioural oestrus was generally terminated as mating, between 5 and 60 minutes from its onset (i.e. the time at which overt oestrus first became apparent). The potential duration of oestrus was not investigated, as no attempt was made to prevent copulation. However, during the 1984 breeding season, one Doe was observed to exhibit induced oestrus for at least eight hours, as the buck, having already copulated with several Does earlier in the day, repeatedly failed to attain a copulatory mount. The observation was discontinued due to the onset of darkness and the Doe failed to conceive at the induced oestrus.

EFFECT OF PRE-PUBERTAL CASTRATION ON GROWTH

Castration of bucks may be a useful tool for providing suitable slaughter animals when entires are normally difficult to handle due to aggression (eg. post-puberty). The growth between birth and three years of age of 21 entire and 11 castrate bucks is represented in Figure 12. Castration was performed using rubber rings at 5 months of age. The two groups were grazed together for almost the entire duration of the trial.

By 15 months of age, the castrates were 9.8% lighter than the entires. The gap increased to 14.0% by 27 months and 20.5% by 39 months.

There was a tendancy for the entires and castrates to show similar fluctuation in liveweight for age. It is not possible at this stage to separate the effects of feed availability, peer pressure and castration upon liveweight. Trials are presently underway to measure feed intakes of entire and castrate bucks up to 20 months of age in order to assess the efficiency of growth of both groups.

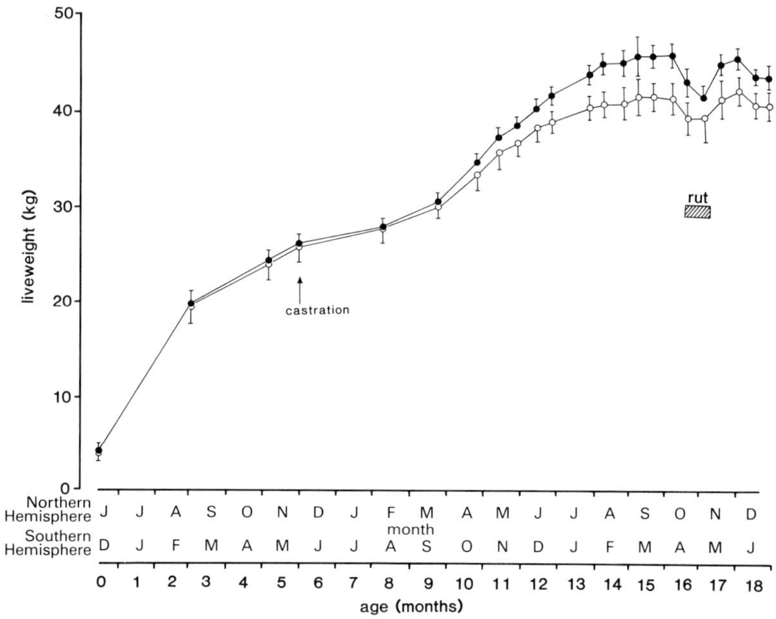

Figure 12

Profiles of mean (and 95% confidence intervals) liveweight, from birth to 18 months of age, for 21 entire (●———●) and 11 castrated (○———○) fallow bucks).

(D) CONTROL OF REPRODUCTION

EFFECT OF THE TIMING OF BUCK INTRODUCTION ON THE ONSET OF OESTRUS IN DOES

The "ram effect" is well documented in the literature. Ewes can be induced to cycle earlier in the season by joining with a ram after a short period of ewe sensitisation to ram absence.

A possible "buck effect" was investigated by introducing single bucks to three widely spaced doe groups at different pre-rut times (ie. 15 March, 25 March & 4 April respectively) (USA—Sept., Oct.) after a minimum period of 30 days in which there was no close Doe contact with adult (>15 months) bucks.

There was no apparent difference in the onset of oestrous activity nor was there a significant ($P > 0.1$) difference in the mean date of first oestrous between groups. This suggests that bucks may not be influential upon the onset of oestrous activity, however, the role of pheromones cannot be ruled out because of their potent nature. It is possible that few farms are capable of removing bucks from the presence of Does at critical times.

OESTROUS SYNCHRONISATION

First oestrus of fallow Does is naturally synchronised over a two week period. However, a greater degree of synchrony may be occasionally required on farms (eg. for artificial insemination).

A 14 day treatment with 12% progesterone CIDRs (Alex Harvey Industries, NZ) inserted intra-vaginally is sufficient to synchronise first oestrous *within the rut.* In one trial with 6 Does, oestrous was tightly grouped between 48 and 56 hours after CIDR removal on 30 April (USA—Oct.). All animals cycled exactly 21 days later, therefore the induced oestrous had resulted in an ovulation and viable corpus luteum.

INDUCTION OF EARLIER OESTRUS

Induction of earlier oestrus and ovulation in fallow Does has been attempted using both PMSG and GnRH treatment after pre-treatment with 12% progesterone CIDRs.

Treatment with 500 iu PMSG (Folligon) upon CIDR removal three weeks before the start of natural oestrous resulted in induced oestrous in 20/21 Does between 48 and 60 hours. Laparoscopic examination at ten days revealed ovulation ratès ranging from 1–4. Conception at induced oestrous was low (2/20 Does) and may have been partly due to poor buck libido observed during the trial. The returns-to-service occurred between days 21–25 after first oestrous, showing considerably more variation in cycle length than untreated animals. This was presumably due to multiple corpora lutea from the induced ovulations. Further studies with PMSG will need to examine lower dosage rates in order to prevent multiple ovulations.

Treatment with low dose, 7-day infusion of GnRH via s.c. osmotic minipumps, (Alzet Corp) following CIDR removal, resulted in oestrous and ovulation in 7/14 Does, as early as mid-Feburary. (USA-August) Oestrus occurred between days 2–5 after CIDR removal. Only one Doe conceived and produced a viable fawn in October (USA-April), at least six weeks before the onset of natural fawning. The technique was very expensive and requires a careful investigation into treatment of bucks to ensure libido and fertility.

CONTROL OF REPRODUCTIVE SEASONALITY OF BUCKS

By comparison with Does, the annual sexual development of bucks is a gradual process lasting several months. Therefore, short term treatments with reproductive hormones are likely to have little effect upon advancement of reproductive competence.

Longer term treatment of seasonally breeding male deer with the pineal hormone, melatonin, is being investigated. A trial was conducted with mature fallow bucks in which melatonin was incorporated into feed and orally administered daily in early summer to simulate short-day length. A 45 day treatment period from 1 December to 14 January (USA—June–July) resulted in a positive, but somewhat transitory, shift in several reproductive characteristics.

The results suggested that for this form of melatonin treatment to be fully effective, bucks need to be treated later in the season and, perhaps, for a longer period. Further trials are continuing.

SEMEN COLLECTION

Semen was collected from fallow bucks by electro-ejaculation. Between 0.1–1.0 ml of semen was obtained from fully anaesthetised animals, with best quality semen taken between March and September (USA—Sept.–March) (refer Figure 8). Techniques of dilution, storage and thawing have not been evaluated.

POSSIBLE ROLE OF ARTIFICIAL INSEMINATION (AI)

The potential benefits of AI in fallow deer farming are primarily related to the importation of semen from exotic sources to introduce genetically different lines.

Two aspects of artificial breeding have been investigated to date; oestrous synchronisation and semen collection. Further studies are required on semen processing and insemination techniques before AI can be realistically considered as a farming tool. (G. Asher NZ, 1985. 107–125.)

REFERENCES

Armstrong, N., Chaplin, R. E., Chapman, D. I. & Smith, B. 1969. Observations on the reproduction of female wild and park fallow deer (*Dama dama*) in southern England. *J. Zool., Lond. 158*: 27–37.

Asdell, S. A. 1964. *Patterns of Mammalian Reproduction*. Constable and Co. Ltd, London.

Asher, G. W., Adam, J. L., Langridge, M. & Howell, L. J. 1981. Calving and fawning performance of red and fallow deer in NNI regions. *33rd Ruakura Farmers' Conf. 1981*: 97–99.

Asher, G. W. & Adam, J. L. 1985. Reproduction of farmed red and fallow deer in northern New Zealand. *Biology of Deer Production*, 1985. Ed. P. F. Fennessy & K. R. Drew. Royal Society of NZ Bulletin No. 22, Wellington (in press).

Asher, G. W., 1985. Proceedings of a Deer Course for Veterinarians. Ashburton, NZ *Reproduction of Farmed Fallow Deer (Dama Dama L.)* pp. 107–123.

Baker, K. 1973. Reproductive biology of fallow deer (*Dama dama*) in the Blue Mountains of New Zealand. Unpublished MSc. thesis. University of Otago, NZ.

Chaplin, R. E. & White, R. W. 1970. The sexual cycle and associated behaviour patterns in the fallow deer. *Deer 2:* 561–565.

Chaplin, R. E. & White, R. W. 1970. The influence of age and season on the activity of the testes and epididymides of the fallow deer, *Dama dama. J. Reprod. Fert. 30:* 361–369.

Chapman, D. I. & Chapman, N. 1969. Observations on the biology of fallow deer (*Dama dama*) in Epping Forest, Essex, England. *Biological Conservation 2(1):* 55–62.

Chapman, D. I. & Chapman, N, E, 1970. Preliminary observations on the reproductive cycle of male Fallow deer (*Dama dama* L.) *J. Reprod. Fert. 21:* 1–8.

Chapman, D. I. & Chapman, N. 1975. *Fallow Deer: their history, distribution and biology.* Terence Dalton Ltd.

Chapman, D. I. & Chapman, N. 1980. Morphology of the male accessory organ of reproduction of immature fallow deer (*Dama dama* L.) with particular reference to puberty and antler development. *Acta. anat. 108:* 51–59.

Couchman, R. C. 1980. *Deer Production in Australia* SCA Technical Report Series No. 5, Canberra 1980.

Cowan, I. McT. 1965. Life and times of the coast black-tailed deer. In: *The Deer of North Africa*. Taylor, W. P. (ed): 523:617.

Eaton, D. 1980. Factors affecting the behaviour and reproductive cycle of fallow deer. *Ratel 7(3):* 5–8.

Goss, R. J. 1983. *Deer Antlers: Regeneration, Function and Evolution*. Academic Press, New York.

Guinness, F., Lincoln, G. A., & Short, R. V. 1971. The reproductive cycle of the female red deer, *Cervus elaphus* L. *J. Reprod. Fert. 27:* 427–438.

Hamilton, W. J., Harrison, R. J. & Young, B. A. 1960. Aspects of placentation in certain cervidae. *J. Anat. Lond. 94:* 1–33.

Harrison, R. J. & Hyett, A. R. 1954. The development and growth of the placentomes in the fallow deer (*Dama dama* L.). *J. Anat. Lond. 88:* 338–355.

Jasczewski, J. Z. 1954. The effect of changes in length of daylight on the growth of antlers in the deer (*Cervus elaphus.* L.). *Folia biol., Praha 2:* 133.

McFarlane, D. 1965. Perinatal lamb losses. 1. An autopsy method for the investigation of perinatal losses. *NZ Vet. J. 13:* 116–135.

Rambotti, F. 1985. Deer farming in Italy. *Biology of Deer Production,* 1985. Ed. P. F. Fennessy & K. R. Drew. Royal Society of NZ Bulletin No. 22, Wellington (in press).

Reinken, G. 1977. Use of meadow and heathland by fallow deer. Translation of German manuscript.

Ueckermann, E. & Hansen, P. 1968. *Das Damwild.* Verlag Paul Parey, Hamburg-Berlin.

Problems at birth are an exception to the rule. The fawn shown here survived and lived. The mother was a very small two-year old, probably below 66 pounds at breeding season and most likely a late fawn herself.

Photo: Georg Rüssmann, West Germany

16

Fawning

Fawn survival is the key to successful deer farming, and since good preparation can mean the difference between success and failure for the year, it is worthwhile to give careful consideration to any means of reducing fawn mortality.

As we have said before, good management dictates that the precautions for fawn survival begin during the mating season. If the sires and Does are in good condition and the Does carry appropriate weight, you have a head start on the problem. Birth rate depends on the condition of the female. Sub-adult females conceive late or not at all when they are kept under poor feeding conditions. During the gestation period, there are several preparations that will help protect the newborn fawn.

1) Well-planned fencing, made fawnproof by extra wires or special netting, will prevent escape and also keep the new-borns from becoming entangled. Fawn deaths due to fence hangings could cause heavy losses.

2) Prepare the pasture for a high percentage of green grass, clover, and alfalfa. If the pasture should require mowing in the late spring, you must rotate to another pasture to avoid mowing over hidden fawns.

Also, the fetus' main development is during the spring, when the grass grows, so pasture preparation is doubly important!

3) Keep watch for aborted fetuses and have them examined for possible determination of the cause. A small fawn crop may be prevented by early detection of disease and proper treatment. Poor nutrition of the pregnant Doe may be another cause.

4) Try to keep the mating season within October, November, and the first half of December, to prevent late fawns, which cause problems at weaning time.

5) Dedicated attention to good nutrition for pregnant Does is likely to increase the birth weight of fawns. Low birth weight greatly reduces

the fawn's chances for survival. Fawns below 6½ pounds (3.0 kilograms) have only a 50% chance of survival. (The average is 8.6 to 10 pounds for males and 6.6 to 9 pounds for females. Young Does tend to have lighter fawns and, therefore, a higher mortality rate among offspring.

Underweight fawns can be recognized by a deficient length of hind legs.

APPEARANCE AND BEHAVIOR OF NEWBORN FAWNS

The majority of fawns are born from early June to late July, but if you have taken your bucks out in late December, fawns can be born well into August, which, as we have pointed out, can create problems. Try to get the bucks out as early as possible, preferably by the middle of December.

A single fawn is the rule, but twins do occur, although they are hard to substantiate, because fawns will suckle from Does other than their own mother. As many as four to five fawns may try to suckle from one Doe.

The Amercian biologist Margaret Altmann has observed a kind of kindergarten system with wapiti deer, and I have seen the same thing with fallow deer. The fawns stay with two or three Does while their mothers go out to pasture. I have seen up to a dozen or more with their "babysitters." Fawns will be kept hidden for the first days of life, during which they will have no discernable scent, in order not to attract predators. After several days they begin to follow their mothers.

The tips of a fawn's hooves have a soft, waxy-looking cover that prevents injuries to the mother during gestation and birth. After a few days, the substance shrivels up to expose the sharp tips of the hooves.

If there is an attractive area outside of the pasture, such as tall grass in which the fawn could try to seek shelter, it will attempt escape, so it is a good idea not to provide any temptations directly around the outskirts of the pasture. At Lucky Star, we have arranged with our New Zealand fence manufacturer for an additional horizontal wire in the fence to prevent fawns from escaping. This new fence with 14 line wires is now available for deer farmers. (See FENCING.) Although we feel it is best to leave the deer alone during fawning, the layout of some farms may make it a good idea to do a "walk around," to check for injured or dead fawns. If any are found, they should be submitted for a post-mortem examination to determine the causes, because a proper diagnosis and treatment plan could save the lives of many fawns in the event of an outbreak of disease. The walk around is important if you have a problem with predators, and you should accustom the deer to it PRIOR to birth.

BIRTH WEIGHT AND WEANING WEIGHT

The average weight of male fawns at birth is 8.6 to 10 pounds, with females weighing approximately 10% less. Under the proper conditions fawns will gain an average of 2.2–2.8 pounds per week for the first four months.

It is important to keep an eye on the fawn growth rate, to meet the ideal average weaning weight of 40–55 pounds. After weaning, the growth rate normally slows to 1 pound per week until winter, when it slows to ½ pound per week. In the spring and summer of the first year, age 9–12 months, bucks will grow at almost twice the rate of Does, or approximately 2 pounds per week to 1 pound for Does.

HAND REARING OF FAWNS

Circumstances beyond control, such as unintended human interference or Doe rejection, or low birth weight, may lead to a situation where the fawn must be reared by humans or it will perish. Some farmers have been successful at rearing mother-rejected or underweight fawns. In general, the animal must be kept warm and clean. Wipe under its tail with a damp, warm cloth to stimulate bowel movement. Also, keep its face clean. It should be fed a mixture of five parts goat milk to one part water (approximately 120 milliliters) by getting it to lap from a bowl or take from a bottle with a lamb teat.

The German deer farmer Josef Rüssmann has discovered that the chances for success in hand-rearing fawns is greatly improved if you provide some free-choice topsoil. As the fawn becomes a ruminant, the size and bacterial content of its four compartment stomach changes. The ingestion of topsoil seems to provide some necessary elements for that change.

Fawns will be kept hidden so well for the first days of their life that it is hard to see them.

Photo: Author

17

Weaning

Why wean at all? First, it gives us the chance to give better care to the fawns, and they are the moneymakers. Especially in areas with harsh winters, or a cold wet autumn, weaning has definite advantages. One good reason to wean is the ability to give feed priority to young stock. The fawns are the most important variable in deer farming and we should, therefore, pay special attention to them.

Another consideration, especially in our area, is the weather. If we wean our fawns in early October, before the rut, we will probably save more fawns from bad weather than we would lose because of young age. Which means that one must provide good shelter after weaning. Fawns conserving metabolism is so high that they might not make it through the winter without dry shelter.

Early weaning is very important for single sire mating. If Does were kept in large herds before, this is the opportunity to split them into special mating groups. Weaning before the rutting season allows the Does to increase liveweight and level of condition, if there is enough feed available. If Does ovulate while still lactating, the estrus cycle may be suppressed. High Doe numbers can create stress mortality in fawns, and keeping them with their mothers during rutting season creates all kinds of problems for the fawns. Early weaning also helps to control worms and parasites, by enabling one periodically to worm the fawns and shift them to "clean" pastures.

If possible, wean smaller mobs to avoid stress related mortality. Any pre-weaning stress should be minimized. Do not chase the deer into yards, do not bring them long distances, and do not hold them in the yards or working areas too long. It might even be a good idea to wean fawns first and do all the other jobs, such as tagging, tattooing, injecting, etcetera, later.

Yard pens should not allow weaned fawns to see their mothers or other neighboring fawn groups. They might attempt to traverse fences and could injure themselves. It would help if the weaning paddock were one or two paddocks away from the others.

Avoid holding the fawns in the yards too long after weaning, because fawn growth rates could be reduced. They will settle down just as well in their own paddock, which should be dry and have shelter available. Always bear in mind that the development of the kidney fat capsule (*capsula adiposa*) is extremely important to the survival of fawns and should have developed by this time. If you happen to lose some fawns, have your veterinarian perform a post-mortem examination and check the *capsula adiposa*.

18

Dehorning and Castration

CASTRATION

Our experiences at Lucky Star have given us a negative outlook on castration. Some years ago, we did a test with twelve male fawns, each of which weighed about 45 lbs. at the age of 3–3½ months.

For this test:

3 fawns were castrated
3 fawns received Ralgro implants
3 fawns were castrated and implanted with Ralgro
3 fawns were kept as a control group

At the age of fifteen months, these yearlings, identified by special eartags, were harvested and weighed along with the others.

1. The three castrated yearlings were among the smallest, and we estimated that they had a higher percentage of fat than the others.
2. The three yearlings with implants were not noticably heavier than the others. One had died during the summer.
3. The three yearlings which were castrated and implanted with Ralgro were the heaviest, but only by 2 or 3 pounds One of that group had also died during the year.
4. The three in the control group were, in our overall opinion, as good as any of the others.

This test convinced me that it is better to keep the animals as natural as possible. They do very well on natural feed, and nothing else seems to be required.

DEHORNING

Dehorning is a different problem entirely. Experts around the world continue to debate the pros and cons of dehorning. According to Prof. Dr. Hermann Bogner from Bavaria, West Germany, dehorning is equivalent to a "psychological castration". Antlers are not only weapons that show strength and gain respect, they are the animal's defense. Prof. Bogner agrees that antlers pose a certain danger to other animals and humans, but fallow bucks, unlike red deer stags, have not yet killed anyone. They will, however, kill their rivals during the rut. But if one keeps bucks of different age groups in a pasture, instead of rivals, there is very little danger. According to Prof. Bogner, there is generally no justification for dehorning bucks.

Contrary to that opinion is the view of Prof. Dr. Günter Reinken, from the Landwirtschaftskammer Rheinland, in Bonn, West Germany. He thinks that dehorning is an absolute necessity. He argues that polling, or permanent dehorning, of four- to six-month old male fawns with an electric dehorning device is probably the best solution. It has absolutely no effect on fertility and rut behavior other than to keep the animals quieter and not so aggressive during the rut. There have been very few reported injuries in herds where all the bucks have been dehorned. The pedicle on which the antler grows can be easily burned under anesthetic and the growth of antlers prevented permanently.

According to the "New Zealand Fallow Deer Farmers Discussion Forum" in 1985, yearling bucks need to be de-antlered before slaughtering. Older bucks can be de-antlered either above the coronet, allowing the antler to grow again the next year, or below the coronet under anesthesia. Dehorning, the New Zealand farmers agree, reduces aggressiveness and damage from fighting. Dehorned bucks suffer no loss of fertility and successfully service Does.

Dehorning is definitely an area of deep concern to us, since we are losing bucks every year during the rut because of injuries from fighting. Although one could say it is best to keep the animals in their natural state, with antlers, it should be considered that they are kept in an unnatural environment to begin with, and some adjustment to accomodate that fact may have to be made. At this point, we are planning to dehorn some of our best male fawns and keep them for breeding purposes, and we will see what difference this makes.

I think each deer farmer will have to decide for himself how he wants to handle it. Both arguments have something favorable. Why shouldn't the strongest bucks fight to breed the most Does? Antlers help the strong animals to gain respect. For some, aesthetical considerations tip the scales in favor of antlered bucks.

On the other hand, it is economically worthwhile to consider permanent dehorning to control serious injuries from fighting. Who can afford to lose several bucks every year?

This is an area where the dicussion will probably continue for some time.

INHIBITION OF ANTLER DEVELOPMENT OF FALLOW BUCKS: THE POLLING TECHNIQUE

Dr. G. W. Asher

Ruakura Animal Research Station, Hamilton, New Zealand

[*Reprinted with kind permission of the author*]

Most fallow deer farmers are fully aware of the problems encountered with the management of antlered bucks. Even deer that have only vestiges of hard antler (i.e. buttons or spikes) are capable of inflicting severe injuries to other deer, especially during yarding. At such times, normal herd hierarchy is disrupted and bucks become unusually aggressive towards other members of the species. The resulting injuries probably account for more adult fallow deaths on farms than any other single cause.

As bucks do not exhibit overt signs of aggression during the annual velvet antler phase they can be safely yarded between October and January (USA—Apr.–July). However, many farmers will not attempt to yard bucks at other times of the year and this can place major constraints on intensive herd management, such as pre-rut sorting of sires for joining with does and the subsequent removal or replacement of sires at the end of the rut.

This problem prompted an investigation into the possibility of permanently inhibiting pedicle and antler development of fallow bucks, especially in individuals retained as herd sires.

The removal of pedicles from mature bucks (i.e. older than one year) does *not* completely inhibit antler growth, although it may be a useful practice for short-term handling requirements (e.g. transport of pubertal bucks to the Deer Slaughter Premise). On the other hand, "polling" bucks as weaners, by cauterising the primodial pedicles with an electric disbudding iron (as used for dehorning calves), appears to inhibit completely subsequent growth of both antlers and pedicles in *most* individuals. The success of the procedure depends upon the degree of primordial tissue (periosteum) destruction, and this is achieved with considerably less trauma on five-month-old bucks (at approximately 25 kg liveweight) for which pedicle primordia are palpable but not necessarily visible, than for older bucks with clearly visible pedicles. Furthermore, as most weaner bucks will eventually be slaughtered before the onset of aggressive behaviour at 15 months of age, only those weaners intended for use as sires need be treated. It is likely that the top 20% of weaner bucks (in terms of liveweight or any other selection criteria) can be selected at or close to weaning. These are the animals that should be kept as potential sires and are the likely candidates for polling.

The following polling procedure is intended for use by veterinarians as it is a reasonably skilled operation that, for ethical reasons, requires the use of local anaesthetics. A similar technique has been used with considerable success on German fallow deer farms (Reinken 1977).

POLLING PROCEDURE

Draft individual bucks for polling in May (USA—Nov.) (five months of age) or at 25 kg liveweight.

Pre-heat the disbudding iron (electric or battery operated calf disbudding iron with 2 cm diameter concave cutting disc).

Clip the hair covering the pedicle site to highlight the primordial pedicles.

Administer a local anaesthetic to the general region of both primordial pedicles.

After allowing time for the local anaesthetic to take effect, place the concave disc of the hot disbudding iron directly over each of the pedicles in turn, apply pressure while rotating the disc slightly and remove the resulting wad of skin to expose the bony structure of the developing pedicle.

With the cutting edge of the disc, remove the bony pedicle structure flush with, or even slightly below the level of, the frontal bone; making sure to cut away from the animal's eye.

Complete the operation by cauterising the general area over a 1.5 cm radius. It is important to eliminate all primordial pedicle tissue at this stage, otherwise small antlers may develop later in life.

No further treatment is required, however, it may be wise to vaccinate fawns against clostridial diseases several weeks prior to treatment.

Success rate: Of 29 weaner bucks treated in either 1981 or 1982, three developed a small weak antler on one side by four years of age. The other bucks have remained polled.

Performance: Polled bucks have been used as sires for the last three breeding seasons. Trials have been conducted in which Does and polled bucks have been yarded together up to 80 times during the mating period. During mustering and yarding the bucks did not exhibit aggression towards the does. Needless to say, the successful outcome of these trials was partly dependent upon the ability to yard deer frequently without aggression related problems. To this end, the polled bucks were worth their weight in gold.

However, polling does not completely inhibit aggression in fallow bucks. Rival polled sires will frequently fight during the rut if they share close territorial boundaries. Furthermore, adult polled bucks in bachelor groups also will have sparring matches during the autumn and winter months. The head to head confrontation of two polled bucks is an amusing spectacle to observe, as the forward momentum often causes one buck to complete straddle its opponent in the absence of locking antlers.

Frequent sparring generally results in little more than temporary, partial baldness. However, in 1985 we experienced some deaths in our bachelor group of polled bucks that were directly *attributable to fighting* in the paddock even though they showed little aggression when yarded together. This problem was alleviated by breaking down the bachelor herd structure into buck pairs (in the absence of a large, broken paddock on the Ruakura Deer Unit) for the remainder of winter, as it appeared that groups of bucks were "ganging up" on individuals. This was probably induced by a high stocking rate within small (0.25 ha) paddocks and the fact that all the bucks were of the same age and approximate liveweight, leading to a fragile hierarchal structure. It is highly likely that, given the same conditions, the same problem would have occurred with antlered and velveted bucks.

RECOMMENDATIONS FOR THE MANAGEMENT OF POLLED SIRES

Single sire mating of small (<35) numbers of Does in each group is probably preferable to multi-sire mating with larger groups of Does. Regardless of antler status, this tends to reduce valuable energy being lost defending territories. Also it ensures that top bucks (i.e. selected by merit) do actually service Does and sire offspring.

If multi-sire mating is practised, ensure that each buck has an adequate area for maintaining a defendable territory that does not overlap with that of other bucks.

Never run polled *and* antlered bucks together, especially during the rut.

If bucks (polled or antlered) are to be removed from mating groups at the end of the main rutting period (i.e. May–June) (USA—Nov.–Dec.), maintain them in loose bachelor herds; do not force them into high density herds over winter. It may be pertinent to overwinter them in large broken (e.g. some bush cover) paddocks. However, still provide high quality feed as their body reserves of fat are depleted after the rut.

Present experience indicates that yarding of Does in the presence of a polled sire buck presents few problems. However, if a particular buck becomes unusually aggressive it may be wise to eliminate him as a future sire prospect. Even so, it is unlikely that an aggressive polled buck will do as much damage as a buck bearing antlers or vestiges of antlers.

Conclusions: Polling is not the complete answer to eliminating buck aggression problems on the farm. However, it may go a long way to reducing stock injuries during yarding, and as such deserves at least some consideration as a management tool. The success of the technique in permanently eliminating antler growth has yet to be fully evaluated but present indications are that, providing the operation is performed properly, antler growth is eliminated until at least four years of age. If the practice is to be adopted, the polling operation should be performed under veterinary supervision and will be no worse than velvet antler removal.(Asher, Dr. G. 1986. Proceedings of a Deer Course for Veterinarians New Zealand Vet Assoc., Rotorura, July 1986. pp. 202–206.)

REFERENCES

Reinken, G. (1977) Grün-und Brachlandnutzung durch Damtiere. Angregungen fur Produktion und Absatz, 10.

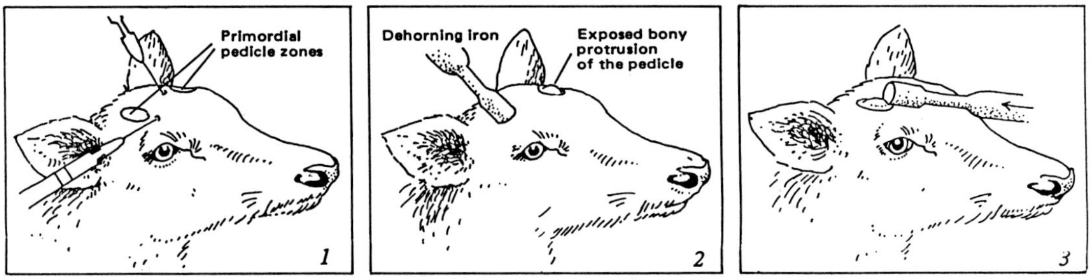

Courtesy of THE DEER FARMER *(Deer Farmer Publications Ltd.) New Zealand*

19

Health and Veterinary Management

My grateful thanks to an editor friend for technical comments. Drs. Brunner, Dubovi, Kallfelz, & Lein (Cornell/NYS Coll of Vet Med) for their constant support. The Kerckerincks, Mr. Bruno, and the whole LSR staff for unwavering determination and my Husband, Son, & Secretary for their patience.

M. W. Giaquinto, DVM

HEALTH AND VETERINARY MANAGEMENT

GENERAL INTRODUCTION

The domestication of each species has been and is a complex evolutionary continuum. The biology of each individual species has mapped the course for transition from feral to domestic state. Husbandry and management are, therefore, a process of fine-tuning the new "ecosystem" of animal, man, and altered habitat.

Wolf and jackal-type dogs met man (Konrad Lorenz) and their various relationships with man, dependent upon breed or type, became legion. Similarly, the species of pleasure, production, and work animals each developed a unique relationship with man observant of interdependent needs.

Archeological bone finds indicate that Mesolithic and Neolithic man systematically harvested venison in much of the world millenia ago. The Romans enclosed deer, keeping them in parks for pleasure, or ranching them for profit. Columella 65 A.D. "Wild creatures, such as red, roe and fallow deer .... sometimes serve to enhance the splendor and pleasure of their owners, and sometimes to bring profit and revenue. Those who keep game shut up for their own pleasure are content to construct a park .... in the neighborhood of the farm buildings, and always give them food and water by hand. Those on the other hand who look for profit and revenue, when there is a wood near the farm (for it is important that it should not be far out of sight of the owner) reserve it without hesitation for the above mentioned animals." (15. Fletcher, 45)

Physiological constraints have limited the intimacy of the deer/man relationship. The flighty temperament, mechanical ability to fly through the air, antler cycle, rutting behavior, and lack of need for constant close handling have kept deer herds at some distance from their herders most of the time. Unlike the dairy cow or goat, whose role requires handling and milking twice daily, or the sheep whose nature and needs have led to close familiarity with man, both farming and ranching of deer dictate measures of close care and benign distant supervision.

What biological and physiological considerations bound any thoughtful deer health, husbandry, and management program? Answering this question has inspired much observation and study over the centuries. In order to know the abnormal, one must first know the normal. Subscribing to this principle, man has observed and recorded the actions, habits, and husbandry of deer at least since the 16th century. The French publication in 1561 of Jacques du Fouilloux's **La Venerie et l'Adolescence** (essaying the habits of wild red deer) was a best seller, being reprinted many times and then translated into several other languages. The English version, translated by George Tubervile in 1575, was titled **The Noble Arte of Venerie or Hunting:**

"Wherein is handled and set out the Vertues, Nature, and Properties of fivetene sundrie Chaces togither, with the order and maner how to Hunte and kill every one of them. Translated and collected for the pleasure of all Noblemen and Gentlemen, out of the best approved Authors, which have written any thing concerning the same: And reduced into such order and proper termes as are used here, in this noble Realme of England." (7. Short, 461–2)

Over the centuries, man has perpetuated a cycle of deer farming, ranching, hunting, and varying combinations of the three worldwide. Over the last few decades, man has determined a course of intense deer husbandry in numerous parts of the world. Scientific accumulation and compiling of empirical data emerged, followed by more exacting controlled scientific studies. Although the New Zealanders have farmed deer in a serious way only since 1970, they have done much to organize and disseminate the industry's commercial and scientific information worldwide.

The following deer health, herd health, and nutrition information is drawn from the bare bones of personal experience fleshed out by the worldwide deer farming experience. Especially important models for us in North America are the British, the Europeans, and the New Zealanders, the last of whom in less than two decades of intense deer farming have asserted themselves as industry world leaders.

ANESTHESIA

Desired Effects:

Analgesia or painkilling
Muscle relaxation
Sedation or tranquillization

Indications:

Capture
Handling
Surgery (ex. calving, velvetting)

GENERAL ANESTHESIA

Drugs

Rompun (xylazine), an injectable, non-narcotic anesthetic discovered 25 years ago by Bayer chemical company in Germany (available here through Haver, Mobay United States division), is widely used in veterinary medicine in many species. Its ease of administration, wide margin of safety, limited undesirable side effects, and usefulness across species lines has made Rompun a worldwide marketplace success. In the United States, Rompun's on-label uses are for horses, fallow deer, mule deer, sika deer, white-tailed deer, and elk. (Package insert materials. **The Use of Rompun [in deer]**. Haver.)

Experimentally, anesthetic reversal of Rompun is possible with a newly developed alkaloid antagonist, Yohimbine hydrochloride–from the bark of the African tree *Corynthe yohimbe.* Yohimbine alone or in combination with 4-aminopyridine (4-AP), another emerging antagonist, can reduce recovery time from hours to minutes. In obstetrical cases, Yohimbine offers the advantage of immediate post partum resuscitation of the neonatal fawn, whose immature organ function precludes efficient anesthetic metabolism.

Yohimbine and 4-AP also appear to facilitate recovery from anesthesia with a Rompun/Ketaset mixture in use in other countries, but not on label in the United States (studies involving bighorn sheep).

Although all of the anesthetic and reversal drugs mentioned have a high safety index, route of administration is critical. If administered directly into the circulatory system, intravenous (versus intraarterial) route is a must. Intraarterial injection results in immediate central nervous system excitement. Sequellae are irritability or hyperexcitability, sometimes progressing to recumbency, seizure, and death. The carotid artery is superficial and closely associated with the jugular vein in the neck of the deer, so care is required in injection technique.

Rompun may also be administered intramuscularly, an advantage in the capture of deer via dart gun. Sudden deaths can occur due to hyperexcitability-stress-shock syndrome (if the deer have been hotly pursued) or due to chance intraarterial injection. (6. Winter, 1985 van Reenen, 15-6; January, 1986 Renecker/Olsen, 31)

Positioning

Once under general anesthesia, the deer should be positioned to minimize bloating, which can itself lead to respiratory embarrassment and death. Positioning will be dictated to some extent by the procedure to be performed. The ideal position is sitting upright on the haunches — a sheep seat might be helpful here.* The next best positional choices are sternal recumbency or right lateral recumbency. Left lateral recumbency should be avoided. If elective surgery is planned, decreased intake of volatile feeds for 24 to 48 hours presurgery will reduce the risk of bloat. Common sense must be used here — feed deprivation of debilitated animals or animals during times of great stress may itself be life-threatening. (15. Way, 138)

In our experience, unanesthetized fallow deer become calmer once properly positioned upright in a sheep seat. We have performed limited nonsurgical procedures in this manner.* **Caution is required of the operator in positioning the deer!

REGIONAL ANESTHESIA

Drugs

Many regional anesthetic agents exist. They all work by blocking sensory nerve impulses. Injection of one of these solutions locally may be all that is necessary for minor surgical procedures. Local analgesia may also be used in conjunction with general anesthesia, often to reduce the amount of general anesthetic needed.

Technique

Multiple injections of the surgical area may be used, forming **line, U, L, or ring** blocks. Or a limited number of injections may target a specific nerve or group of nerves. (13. Evans, 468)

Notes: *No* meat withdrawal times have been established for food-producing animals.

For more complete information on fallow deer anesthesia, consult your veterinarian.

The Study of Disease Outbreaks in the Animal Population

EPIZOOTIOLOGY

The Diseases of Aniamls Transmissible to Man

ZOONOSES

The Study of Their Human Health Significance

EPIDEMIOLOGY

[*Adapted with kind permission from Mackintosh and Beatson. "Relationships between Diseases of Deer and those of Other Animals," Biology of Deer Production*, 77-82] (7.)

A triad of contributing factors must exist before disease can occur. The host must be susceptible, one or more infective organisms must be present, and the ambient environmental conditions must be favorable to the infective organism and hostile to the host.

SUSCEPTIBLE HOST	+	INFECTIVE AGENT(S)	+	CONDUCIVE ENVIRONENT	= INFECTION
HOST FACTORS		AGENT FACTORS		ENVIRONMENTAL FACTORS	
Sex Age Stress Nutritional status Genetic resistance Acquired resistance Concurrent disease		Virulence Pathogenicity Host specificity Infective dose Ability to multiply Resistance to environmental insults	+	Climate Food availability Population size and density Seasonal factors Ecological relationships	

A classic case of this triad mechanism is shipping fever pneumonia of cattle (which, by the way, is something of a misnomer, as the disease need not be associated with shipping at all). Shipping provides the stressful environment. A different stress may be substituted here—overcrowding, poor ventilation, etcetera. The unvaccinated or undervaccinated bovine is the susceptible host, and the agents are the everpresent, opportunistic bacterial organisms comingled with one or more hot viruses. These produce one nasty etiological effect—pneumonia!

An apt example in deer is malignant catarrhal fever (MCF).

SUSCEPTIBLE DEER	+	MICROBIAL AGENT(S)	+	CONDUCIVE ENVIROMENT	=	INFECTION
Possibly sika and red deer are more susceptible, while fallow deer are considered fairly resistant.		Virus plus secondary bacterial oportunists cause predominantly the head-and-eye form in sporadic cases and the nervous and intestinal form in herd outbreaks.		Factors include: –change of seasons –nutritional stress (winter feeding time) –intensive grouping –proximity of possible nonaffected carrier ruminants: especially sheep, although sometimes cattle.		Death usually occurs within ten days; survivors are unthrifty.

Any disease outbreak will be a grab bag of multiple-host, multiple-agent, and multiple-environmental factors. So complex is the interaction of these elements that the presence and significance of each and every determiner is rarely known. Even in grave situations, in which a complete necropsy or post-mortem examination is performed (with blood analyses, bacterial and viral culturing, and histological tissue section appraisal), a truly definitive diagnosis sometimes cannot be reached. Even a satisfactory presumptive cause of death may not be apparent. As discouraging as this may be, continuing examination of any and all disease occurrences will hopefully, in the long run, reveal the causal agent or agents, contributing factors, and a solution to the problem affecting the stock. This solution may involve changing host susceptibility through biologicals (vaccines) or improved nutrition. It may require environmental changes on the part of management—better housing, decreased population density, precise balancing of the nutritional program. Or it may attack the organism itself through practical hygiene or most recently through direct antigen neutralizing by genetically engineered monoclonal antibodies. This last strategy is in its infancy in both human and veterinary medicine. An example from veterinary medicine is the genetically engineered monoclonal antibody to the *Escherichia coli* bacteria of calf scours. An oral preparation given by mouth to calves at risk, this antibody medicant neutralizes the bacteria (or antigen) before disease can occur. No parallel products exist for deer at this time, but the immunologic technique involved is the way of the future in all branches of medicine.

WAYS IN WHICH DISEASE MANIFESTS ITSELF

ENDEMIC DISEASE: a disease that maintains itself over time in a relatively steady state in a particular group of animals. An example of this is lungworm infection of deer, although fallow deer are considered to be somewhat resistant.

SPORADIC DISEASE: a disease that appears in scattered or isolated instances and then disappears permanently or temporarily. Examples of sporadic disease include: 1)subclinical infections that become clinical periodically like leptospira bacterial infection, which causes abortions in Does and death of young deer at times, or 2) introduced infections that come from contact with other species, for example, equine herpes virus I, which is normally confined to horses but has been reported in fallow deer; or rabies, especially in areas where rabies is endemic in the wildlife population.

EPIDEMIC DISEASE: a disease that is not usually present but which, once present, spreads rapidly until it either runs its course or becomes endemic. Examples include: 1)the foot-and- mouth disease outbreak in California of cloven-footed ungulates (including white-tailed) deer in the 1920's, which was eradicated by a rigorous quarantine, test, and slaughter program*, and 2) bluetongue/epizootic hemorrhagic disease of white-tailed deer outbreaks, which occur periodically.

**More than 130,000 animals were destroyed to halt that epizootic—approximately 22,000 white-tailed deer and 58,000 cattle were slaughtered between 1924 and 1926. Fallow and red deer are not considered effective transmitters of FMD. (2. Cottral & Callis, 157)*

MODES OF TRANSMISSION

Direct/Animal to Animal Transmission
- Aerosol
- Direct Contact
- Oral
- Oral-Fecal

Indirect Transmission (air, feed, soil, water)
- Vehicles
- Fomites (inanimate objects)
- Intermediate Hosts (ex. cattle, sheep, swine, and certain wild ungulates for the canine taeniid tapeworm) (9. 164)
- Invertebrate Vectors (ex. flies, fleas, lice, mosquitos, ticks)

Non-Transmission: The Dead-End Host
- Infection—Death

or
- Infection—Recovery without shedding

Carrier State—Condition of shedding an infective agent either during the course of an asymptomatic infection or during the incubation or convalesence or post-convalesence (chronic carrier stage) of a disease that was clinically apparent.

Long-term Carrier (Resevoir)—Any animate being or inanimate object in which a contagious agent can live and multiply and from which effective transmission can be completed.

Short-term Carrier—The carrier state may be transient or temporary.

Both farmed and wild deer are susceptible to the same spectrum of maladies. There are relative advantages and disadvantages with each life style. For example:

Farmed Deer	ADVANTAGE Balanced Nutrition (Hopefully!)	DISADVANTAGE Dense Population
Wild Deer	DISADVANTAGE Browse and Forage only (Catch as Catch Can!)	ADVANTAGE Wide Range

It has been stated that deer in the wild are less subject to outbreaks of clinical disease than farmed deer. But, under good management systems, farmed deer should be relatively free of disease, clinical or subclinical. Exact statistical comparisons of farmed and wild deer would be difficult to generate and almost impossible to interpret. Skewing of the data would be inevitable due to Mother Nature's natural selection and survival-of-the-fittest regimes for the wild species; and man's genetic manipulations with attendant husbandry, nutritional, and veterinary care programs for the farmed species.

Nevertheless, we know, empirically, many diseases to which farmed and feral deer are subject, as well as some transspecies implications for deer, other domesticated animals, and man.

> "Disease interactions between farmed deer and other animals fall into 4 broad categories:
>
> (i) diseases brought onto the farm with deer captured from the wild; [ex. bluetongue, EHD] [or assembled from other farms; ex. coccidiosis]
>
> (ii) diseases of farmed deer affecting other domestic livestock; [ex. gastrointestinal parasites and lungworms]
>
> (iii) diseases of livestock affecting farmed deer; [ex. brucellosis, dermatophilosis, leptospirosis, MCF]
>
> (iv) diseases of wildlife present on farms affecting farmed deer. [ex. tularemia—rabbits, rabies—bats, etcetera]" (7. Mackintosh and Beatson, 79)

To this list should be added zoonotic diseases, those transmissable between deer and man. They include: anthrax, brucellosis, various parasitic diseases, pasturellosis, leptospirosis, tuberculosis, tularemia, rabies, and Q-fever. The public health implications of feral deer zoonoses most often touch forestry workers, hunters, and trappers; while specific zoonoses of farmed deer put farmers and ranchers, haulers, end product-workers, and veterinarians at risk. *(7. Mackintosh & Beatson, 77-82)*

Most of the information in the disease sections is general veterinary knowledge and has mainly been drawn from the following references:

1. **Aglink**
5. **Deer Course**
6. ***TDF***
7. **Biology of Deer Production**
8. **The Merck Veterinary Manual**
10. **Viral Diseases of Cattle**
11. **Infectious Diseases of Domestic Animals**
15. **The Farming of Deer**

(See **References** at the end ***HEALTH AND VETERINARY MANAGEMENT***.)

Where nuances have been described, they are referenced in place in the text.

BACTERIA

ANTHRAX caused by *Bacillus anthracis,* a spore-forming bacteria of the same family as the clostridiums, causes an acute or peracute febrile disease in virtually all mammals excluding the rat. (Fowl are also resistant.) Alkaline or neutral, calcareous soil types form worldwide anthrax districts or incubator areas, especially during late summer and early fall warm weather, and particularly if rainy. Anthrax districts exist in parts of Africa, Asia, Australia, Europe, and North and South America. In the United States, enzootic areas include parts of the states of Mississippi (southern valley), South Dakota, Nebraska, Iowa, Arkansas, Louisiana, Texas, California, and Minnesota. Sporadic outbreaks have occurred in parts of Vermont, New Jersey, Delaware, Wisconsin, Utah, Nevada, Oregon, Illinois, Indiana, Ohio, Michigan, and small sectors in a number of other areas.

CLINICAL SIGNS

Three forms of the disease exist:

1. Peracute: animal is found dead without any other sign.
2. Acute or subacute: the most common form, animal is pyrexic, has cardiac and respiratory distress, develops subcutaneous edema and hemorrhage (clotting mechanisms malfunction). The spleen enlarges and becomes soft, black, and bruised. Anthrax is also called "splenic fever." The animal usually dies within one to five days. Occasionally, prompt antibiotic plus supportive therapy may save an individual animal.
3. Chronic: (usually only in pigs) animal has local head and neck swelling of lymph nodes.

PREVENTION AND CONTROL

Vaccination of animals in enzootic areas. Anthrax is the first disease for which a vaccine was developed (Pasteur, 1881). Newer, safer vaccines now exist.
Accurate early diagnosis and treatment with penicillin.
Sanitary disposal of infected carcasses.
Careful inspection and disinfection of animal products including hides (salting the hide does not kill the anthrax spore).

BRUCELLOSIS (Bang's disease) is a contagious bacterial disease of animals and man caused by several species of brucella organisms: *Brucella abortus* (cattle), *Brucella suis* (swine), *Brucella melitensis* (sheep and goats), *Brucella ovis* (sheep), and *Brucella canis* (dogs). While each species is usually infected by its own brucella organism, infection can cross species lines. In man, the cause of undulant fever (Brucellosis) is usually *Brucella abortus*. In Alaska, a reindeer and Caribou enzootic exists caused by *Brucella suis* Type 4. *Brucella abortus* enzootics have been diagnosed in some Elk herds in western Wyoming.

CLINICAL SIGNS

Reproductive problems, abortions in females, genital tract abnormalities and sterility in males, neonatal fading and death in fawns.

Septic arthritis with grossly swollen joints.

PREVENTION AND CONTROL

Test and slaughter programs exist for cattle. This is impractical in Alaskan reindeer and caribou or Wyoming elk.

Vaccination programs may find a place in the management of problem herds. (7. 79, 97, 101)

The clostridial bacterial diseases include: **BLACKLEG, ENTEROTOXEMIA** (overeating disease), **MALIGNANT EDEMA, BACILLARY HEMOGLOBINURIA, BLACK DISEASE,** and **TETANUS**. Of these, the first three have been reported in deer, while the others are a potential threat.

CLINICAL SIGNS

For all of the clostridial diseases, acute death may be the only initial sign. After 24 hours, post-mortem examination reveals lesions indicative of the particular clostridial disease.

BLACKLEG caused by *Cl. chauvoei* and sometimes secondarily *Cl. septicum* typically attacks the skeletal muscles, causing swelling, pain, and a spreading emphysema. Death occurs within 12 to 48 hours. Route of exposure is oral or wound entry.

ENTEROTOXEMIA (*Cl. perfringens* Types B, C, D) occurs when this normal commensal bacteria overgrows in the intestinal tract, usually of young thrifty animals on too much rich feed or milk, or as a sequel to an abruptly changed diet at any age. An exotoxin is released followed by diarrhea or bloody diarrhea, convulsions, and other neurologic signs, and death within two days.

MALIGNANT EDEMA (*Cl. septicum*) closely resembles blackleg, although the disease is more frequently associated with injured or traumatized tissue. Unsanitary conditions or techniques during surgical or vaccination procedures as well as injury, either from fighting, fawning, or accidental trauma, provide the organism a neat route of entry.

Bacilliary hemoglobinuria, black disease, and tetanus have been of little or no importance in deer to date.

BACILLARY HEMOGLOBINURIA (*Cl. hemolyticum*) involves an injury (usually fluke migration) to an infected liver and leads to a toxin-induced, acute hemolytic anemia with port-wine-colored urine and jaundice.

BLACK DISEASE (*Cl. novyi*) occurs after fluke migration through an infected liver produces a parasitic hepatitis, and usually acute death.

TETANUS (*Cl. tetani*) results as a sequel to an infected wound. Stiffness, then muscular tetany (lock-jaw) proceed, followed by death in most cases.

PREVENTION AND CONTROL

In deer, multivalent clostridial cattle or sheep vaccine has been administered experimentally. Boosters are desirable: 1) prefawning (hopefully in order to confer colostral immunity from the doe to the nursing fawn); 2) prior to anticipated times of high risk (exposure to lush feeds or drastic changes of diet).

In the event of an outbreak, vaccination plus simultaneous penicillin injection of the herd will stem the tide. Sanitary disposal of carcasses—deep burial under lime, or burning—is important to prevent further dissemination of clostridial spores by predatory carnivores, running water supply, or mechanical transfer. The spores can remain dormant, but viable for decades.

CORYNEBACTERIUM PYOGENES, a normal commensal of skin surfaces, becomes pathogenic when it gains entry to the body. Corynebacterium infections tend to be malodorous, in time develop a cheesy consistency, and are often part of a mixed infection involving other opportunistic bacteria.

CLINICAL SIGNS

In deer, *navel ill and fawn septicemia* involve corynebacteria. Orphaned fawns, whose navels have not been well attended, are prime targets. Once through the navel and into the body, the bacteria will spread through the blood stream (septicemia), often seeding vital organs with microabscesses. Joints may fill with purrelent synovial fluid and become grossly swollen. As a sequel to this problem, animals tend to be unthrifty. But even those who seem to grow well may go to pieces under stress—such as fawning—several years later.

Corynebacteria has been isolated from *foot rot* and *foot abscess* conditions in deer. As in other species, infection is usually mixed, especially involving the other typical fusiform foot rot bacteria.

Abscesses associated with skin traumatized for any reason (fighting, punctures, unsanitary surgical or vaccination materials/methods) generally contain a mixed infection with corynebacteria predominating. *Escherichia coli* and/or *Staphlococcus aureus* are also sometimes present. Abscesses often change from soft and fluctuant containing serosanguinous material to hard walled-off masses of caseous material.

PREVENTION AND CONTROL

Navel ill: Dip navels of orphaned or mismothered fawns in 7% tincture of iodine several times to dry cord. Administer antibiotics for three to five days.

Foot problems: Contributing environmental factors should be eliminated—wet, muddy yards. The ration balance should be checked especially for factors affecting skin integrity.

Trim, medicate, and wrap pathologic feet and give parenteral antibiotics as necessary.

Abscesses: Eliminate occasions of trauma; faulty carpentry or disrepairs, overpopulated yards and pastures.

Treat wounds by flushing and medicating. Lancing may be necessary if an abscess does not drain naturally and balloons. Systemic medication is indicated for a pyrexic animal.

NEONATAL SEPTICEMIA/ENTERITIS

ESCHERICHIA COLI (*E. coli*) bacterial infections usually affect only neonatal animals within the first few days of life.

CLINICAL SIGNS

Three forms of the disease exist.

Coli septicemia: sudden death within 48 hours without diarrhea. Death is due to endotoxic shock and contamination of brain, joints, heart, liver, and spleen.

Enteric colibacillosis: "white scours" during the first two weeks of life, death within four to five days if left untreated.

Enterotoxic colibacillosis: "white scours" during the first week of life, death within three days due to endotoxic shock.

PREVENTION AND CONTROL

Promote fawn consumption of sufficient colostrum (a Doe's first milk) within the first few hours of life. Decrease stocking rates to minimize stress and contamination of the environment with *E.coli* (a fecal commensal, some serotypes are pathogenic).

Salmonella species bacterial infections affect neonatal or stressed animals. **SALMONELLOSIS** may be inapparent in the incubation or convalescent stages. These animals, as well as recovered animals, may be shedders.

CLINICAL SIGNS

Signs are similar to colibacillosis, except that diarrhea is fetid and fluid to pasty, or bloody. Septicemia can occur. Dehydration may lead to nervous signs prior to death. Septic shock with sudden death is the usual manifestation in deer.

PREVENTION AND CONTROL

Hygiene
Stress reduction
Vaccination—prior to time of stress or exposure
(unavailing in the neonate who is usually a victim of passive immune deficiency (colostrum deprived) and would not have time to build up its own vaccinal active immunity)

Treatment is often unrewarding, while recovered animals tend to be unthrifty.

TULAREMIA (*Francisella tularensis*) bacterial infection ("Deer Fly Fever") can cause congested, enlarged, sometimes focally necrotic (white-spotted) liver, spleen, and lymph nodes. Diarrhea or pneumonia, and death ensue in severe cases. Deer are rarely infected, but are a possible source of infection for man through contaminated hides or meat. Rabbits, hares, and rodents are the usual reservoir.

Yersinia pseudotuberculosis bacterial infection is a stress-induced disease of intensely farmed deer. Winter is the peak season for **YERSINIOSIS**, and young animals are at greatest risk. The disease is spread by rabbits, hares, rodents, and birds infected by this common soil contaminant. Deer mortality can be great under conditions of stress.

CLINICAL SIGNS

Hemorrhagic enteritis (with watery, green diarrhea) progresses to dehydration, endotoxic shock, and death in the diseases' usual course. Septicemia and death peracutely or acutely within a few days can occur in the more fulminant form of the disease. Lymph node enlargements, if present, can resemble tuberculosis lesions.

PREVENTION AND CONTROL

Wildlife Carrier Control

Environmental contamination reduction should help to decrease infection rate.

Stress Reduction

Inapparent, immunity-conferring infection occurs in the absence of stress, especially the nutritional and crowd stressing of fawns.

Vaccination

Development of an efficacious bacterin is in the works in New Zealand where yersiniosis is enzootic. (5. Mackintosh, Henderson, & Griffin, 138-45; 7. Mackintosh & Henderson, 159-62)

Fallow deer are considered fairly resistant to yersiniosis.

FOOT INFECTION *FUSOBACTERIUM NECROPHORUM* the most prevalent organism in cattle foot abscess—foot rot—septic arthritis complex, thrives in wet, muddy soil and can affect deer. Introduction to the body is facilitated by any foot injury or nutritional imbalance that disrupts the skin integrity. ***BACTEROIDES NODOSUS***, the sheep foot rot organism, may also be present in cervine foot disease.

RUMINANT DIPHTHERIA/NECROBACILLOSIS Infection of the mouth, pharynx, and larynx with ulcer and diphtheritic membrane formation occurs less commonly.

CLINICAL SIGNS

FOOT INFECTION Inflammation, swelling, cracking with exudate, and mixed infection with other bacteria can lead to arthritis and joint ankylosis (fusion). Septicemia and death can occur in extreme cases of cervine foot infection.

RUMINANT DIPHTHERIA/NECROBACILLOSIS Trauma due to overly coarse feeds allows secondary invasion of damaged mucous membranes by *F. necrophorum*, a normal commensal of the gastrointestinal tract. Once bacterial overgrowth begins, a necrotizing exotoxin causes the typical lesions of necrobacillosis (or ruminant diphtheria).

PREVENTION AND CONTROL

FOOT INFECTION

Decreased stocking density of well-drained pastures.

Improved nutrition.

Treatment: trimming of affected feet, topical and systemic medication.

Strict segregation of sheep and cattle from deer pastures.

RUMINANT DIPHTHERIA/NECROBACILLOSIS

Control the feed quality.

Minimize the presence of sharp plant awns, stalky hay, and very coarse grains.

Treatment: sulfonamide and antibiotic therapy, sanitation of drinking and feeding areas and utensils, temporary feeding of a soft mash, and quarantine of affected animals.

LEPTOSPIRA SPECIES* bacteria usually causes an inapparent infection in deer but can cause abortion, nervous signs, or kidney and liver disease and death. Recovered animals may shed the spirochete from the kidney via the urine. The bacteria thrives in wet areas; surface water, streams, etcetera.

**Leptospira pomona*—pig; *L. canicola*; *L. icterohaemorrhagiae*; *L. grippotyphosa*; *L. hardjo*; *L. ballum* (3. 1981, Fletcher)

CLINICAL SIGNS

Abortion in the second or third trimester, icterus (jaundice), photosensitivity (liver-induced) (5. Bertram, 200-1), hemoglobinuria (bloody urine), nervous signs or, in unusual cases, death can occur suddenly.

PREVENTION AND CONTROL

Fencing off of stagnant wet areas.
Vaccination with multivalent vaccine.

Mycobacterium bovis (cattle) and *M. avium* (bird/poultry), the causative bacteria of **TUBERCULOSIS** in deer can appear in feral or farmed deer, especially if present in local cattle or poultry or if enzootic in the wild population (United Kingdom—badger; New Zealand—opposum).

CLINICAL SIGNS

Oral route of entry in deer is followed by absorption and lymphatic spread to lymph nodes first, and then vital organs (lung, liver, spleen, intestines). Wasting, coughing, formation of cheesy, encapsulated, subcutaneous abscesses ensues. Abscesses resemble those of yersiniosis. Rarely, septic arthritis occurs.

PREVENTION AND CONTROL

For initial diagnosis on a farm: post-mortem examination
Meat inspection and condemnation of affected carcasses
Test and slaughter

Tuberculin Screening Test—in the United States 0.1 ml. of 2 mg./ml. bovine purified protein derivative (PPD) tuberculin intradermally with a 26 G x 3/8″ needle in two sites at the caudal tail fold and vulva lip (females) and anal ring (males) with observation of skin thicknesses at 72 hours.

In the United Kingdom or New Zealand intradermally at a mid-cervical shaved site with 72-hour observation of double skin thickness.

A more accurate, follow-up TB skin test used in the United States and other countries is the comparative cervical test using avian and bovine tuberculin with 72-hour reading of the difference between the two sites.

An accreditation Ministry of Agriculture and Fisheries program is being developed in New Zealand where TB has become a serious problem in farmed deer. Skin testing in NZ has not met with complete success, however, because a few deer have been so riddled with TB that they show no reaction to tuberculin.

JOHNE'S DISEASE (*paratuberculosis*) caused by oral/fecal exposure to *Mycobacterium paratuberculosis* bacteria is primarily a disease of ruminants. Reports in deer are rare. Cases have been reported from zoos, however.

CLINICAL SIGNS

Cachexia, intestinal wall thickening accompanied by chronic diarrhea, lymph node involvement (especially intestinal lymph nodes), and death after a protracted illness are the usual course.

PREVENTION AND CONTROL

Eliminate exposure to infected animals or pastures.

Testing is difficult due to the slow growth of the organism. Fecal cultures have a minimum turn-around time of one and a half months, often with four months' being required for complete results.

Treatment is unavailing.

Deer are also susceptible to all the other occasional bacteria to which other mammals are subject. A complete detailing of each and every possible bacterial infection would be encyclopedic and even more tedious than what you have just read.

General References: 1, 5, 6, 7, 8, 10, 11, 15.

FUNGI

DERMATOPHILOSIS (*cutaneous streptotrichosis*) caused by the fungal actinomycete *Dermatophilus congolensis* affects the skin of domestic and wild herbivores. Lumps and crusts appear. Damage to the velvet is a prominent feature in deer. Systemic treatment is indicated in severe cases, while letting the disease run its course may be the best solution in benign cases.

Pithomyces chartarum fungus if ingested from contaminated pastures causes **FACIAL ECZEMA**, a form of liver-induced toxic photosensitization. This organism thrives during warm, moist summer and fall seasons. Liver damage can be severe enough to cause jaundice and death. In milder cases, sunburning of thin-skinned (face), light-colored areas takes place. The liver, if minimally damaged, can regenerate partially. Pasture spraying and shading of affected animals are recommended. Fallow deer are quite sensitive to this organism. Most reports of this disease, however, are from Australia, New Zealand, and South Africa. (15. van Reenen, 133–4)

RINGWORM (*Trichophyton species*) infection, although rare in deer, does occur. Treatment with systemic or topical medication could be attempted although logistics may be a problem. Solving the probable nutritional basis of the disease would be more practical.

PARASITES

INTERNAL PARASITES OF DEER Most of the deer parasitology literature focuses on lungworm infection, because lungworms cause the greatest parasitic loss in deer. Also, for the most part, dewormers that kill lungworms in deer are efficacious against the other internal nematodes or roundworms of deer. Due to the faster metabolic rate and higher body temperature of deer, only a few cattle anthelminthics (dewormers) are effective against cervine lungworms. Currently, Ivomec (Merck) and Panacur (Hoechst) are two effective dewormers available in the United States. Although used widely in other countries for deer, they are not on label for deer in the United States.

Management techniques such as pasture rotation and off-the-ground feeding (combined with strategically-timed, sequential dewormings) should all help to fight parasitism.

In general, parasitism in conjunction with other stresses (especially nutritional) will be responsible for the majority of losses in deer not on a regular anthelminthic program.

Signs of internal parasitism are ill-thrift, weight loss, rough hair coat, diarrhea sometimes, and in cases of lungworm infection, a soft cough (unlike lungworm-infested cattle, which have a remarkable, hacking cough).

In addition to lungworms, gastrointestinal, and other body system nematodes, deer can also be parasitized internally by several species of trematodes (liver flukes) and cestodes (tapeworms).

The details of all the parasite life cycles, control programs, and treatment regimes are beyond the scope of this book. They are better referenced from a parasitology text or specific parasitology articles recounting studies in deer. (See references.)

The following is simply a listing of internal parasites reported in deer:

LUNGWORM NEMATODES

Dictyocaulus
Elaphostrongylus
Protostrongylus

GASTROENTERIC NEMATODES

ABOMASUM	SMALL INTESTINE	LARGE INTESTINE/CECUM
Haemonchus	Bunostomum	Chabertia
Ostertagia	Capillaria	Oesophagostomum
Rinadia	Cooperia	Trichuris
Skrjabinagia	Nematodirus	
Spiculopteragia	Strongyloides	
Trichostrongylus	Trichostrongylus	

OTHER NEMATODES

Elaeophora (circulatory system—arterial worm)
Parelaphostrongylus (nervous system—meningeal worm)

TREMATODES (liver flukes)

Fasciola
Fascioloides
Paramphistomum
Dicrocoelium

CESTODES (Tapeworms)

Echinococcus
Moniezia
Taenia

EXTERNAL PARASITES OF DEER Deer are also susceptible to attack by the array of ruminant flies and their parasitic larvae (warbles and bots), besides ruminant lice, mites, and ticks.

In general, treatment and control would follow the protocols established for other food-producing ruminants. Strict observation of package insert usage recommendations and meat withdrawal times is imperative.

General Reference: 7. Watson & Charleston, 105-117; 9.; 15. Mason, 130 and van Reenen, 129 & 131–2.

PROTOZOA

ANAPLASMA/PROTOZOA-LIKE RICKETTSIA OF DEER

ANAPLASMOSIS is classified by some taxonomists as a protozoal disease and by others as a virus-like rickettsial disease. The causative organism, *Anaplasma marginale*, parasitizes the edge/margin of red blood cells of domestic and wild ruminants causing fever, anemia, icterus, weakness, and death either acutely or after a chronic course. Microscopic "marginal bodies" take up contrasting stain in whole blood smears. The disease transmits mechanically via flies, mosquitos, and contaminated instruments, or biologically through several tick species. A related organism *Anaplasma centrale* is not pathogenic. In the United States, the gulf and lower plains and western intermountain states and California are endemic areas. Deer carry the infection latently and are an important reservoir for domestic ruminants.

TRYPANOSOMA/PROTOZOA OF DEER

TRYPANOSOMIASIS (*Trypanosoma cervi*), infection with a single-cell, flagellated circulatory system protozoa, occurs in deer populations in Alaska and the contiguous states without apparent clinical signs. It is not known, however, whether infected deer could pose a health risk to man and other domestic animals. (9. 64)

COCCIDIA/PROTOZOA OF DEER

The name **COCCIDIOSIS** is most often ascribed to infection with the gastrointestinal *Eimeria species*. Although eimeria are strictly species-specific, each host species has many parasitic eimeria species to call its own. Eimeria cause nonregenerative destruction of gastrointestinal lining cells with accompanying bloody diarrhea, unthriftiness and, in the worst cases, blood-loss anemia, neurologic signs sometimes, and death. Study of the problem in deer is limited, but empirically high stocking densities coupled with other stresses can lead to serious disease caused by the eimeria coccidial protozoa.

CRYPTOSPORIDIOSIS (*Cryptosporidium species*), a minute intestinal coccidial protozoa, has been associated with diarrhea, specifically colibacillosis in deer. (3. 1985, Bonniwell)

SARCOCYSTOSIS (*Sarcocystis hemionliatrantis*), the muscle cyst coccidia of mule deer, completes its life cycle in dogs and coyotes, its definitive hosts. But it can cause acute cardiac/circulatory muscle disease and death in the deer, its intermediate host. Other species of sarcocystis affect deer in North America and worldwide with less dramatic effects. Damage to the carcass/meat end product makes control of canines in endemic deer farming areas important. Canine predatory habits also reinforce the need for strict outlawing of dogs in farmed deer pastures.

TOXOPLASMOSIS (*Toxoplasma gondii*), primarily an intestinal coccidia of the cat, its definitive host, can cause clinical extraintestinal disease in the cat. In its intermediate hosts, all mammals including deer, toxoplasma will manifest itself in either the intestinal or extraintestinal form, too. In both definitive and intermediate hosts, transmission is usually via fecal/oral contamination although transplacental infection can occur. (This is a cause of great concern for pregnant women who attend cat litter boxes.) Empirically and experimentally, deer appear not to show clinical disease. Possible human ingestion of encysted toxoplasma organisms in deer carcasses forms a potential public health threat. At this time, toxoplasma contamination of deer carcasses is considered to contribute little to human infection. Control by elimination of cats (domestic, feral, and wild) from deer environments would be a difficult, if not impossible, task.

VIRUSES

BLUETONGUE or "sore muzzle" is an insect-borne viral disease of domestic and wild animals. The virus (an *orbi*virus or ring virus) undergoes biological transmission through the biting insects *Culicoides species* or midges/"no-see-ums." Cattle and wild ruminants act as reservoir hosts for the susceptible species—most notably, sheep.

The disease is not directly contagious, but relies on the no-see-um arthropod vector for spread. Incidence in North America is spotty and the disease is milder than that seen on the African continent, where it is of most clinical importance and is probably caused by a different viral strain.

CLINICAL SIGNS

In sheep, after an incubation period usually of five to ten days, clinical signs of respiratory distress develop with open-mouth breathing, inflammatory reddening of the face, lips, and ear pinnae. Generalized inflammation of the oral cavity (with ulceration and necrosis) especially involves the dental pad and,in severe cases, the end result is a swollen, cyanotic (blue) tongue.

Enteritis and inflammatory lameness (laminitis) may complete this picture of ill-health. The death rate is high.

Clinical signs of infection in other species may or may not be apparent, and only rarely mimic or become as severe as the sheep signs. Cattle are usually inapparently infected, although pregnant cows may abort or deliver congenitally defective calves. DEER, IF THEY BECOME CLINICALLY AFFECTED AT ALL, WILL MOST OFTEN SUFFER A SUDDEN FATAL HEMORRHAGIC FORM OF THE DISEASE. (See *MISCELLANEOUS PROBLEMS* **STRESS-GASTROENTERIC-HEMORRHAGIC-SHOCK SYNDROME.**)

PREVENTION AND CONTROL

A sheep vaccine is available. Reduction of contact with the vector by insect control measures will decrease disease incidence.

TREATMENT

Symptomatic or none.

EPIZOOTIC HEMORRHAGIC DISEASE OF DEER (EHD), once called "black tongue" in the southern United States, is a viral disease of white-tailed deer, mule deer, and pronghorn antelope. The EHD virus has almost conclusively been proven to be the same as the bluetongue virus, an orbivirus, with the same mode of transmission.

CLINICAL SIGNS

The disease has a high morbidity and mortality in susceptible deer populations. Clinical signs point to a peracute hemorrhagic event with generalized edema, hemorrhage, and cardiovascular collapse. Oral lesions, lameness, muzzle crusting, tongue discoloration, and sometimes even blindness can occur, much like bluetongue in sheep.

PREVENTION AND CONTROL

No vaccine. Insect control as for bluetongue.

TREATMENT

Symptomatic or none, as for bluetongue.

FOOT-AND-MOUTH DISEASE or aphthous fever is a highly contagious viral disease (picornavirus) of cloven-footed animals. Horses are not susceptible. The disease is exotic to the United States since eradication in the early 1930's. Historically, nine epizootics occurred in the United States between 1870 and 1932. In the last United States epizootic, more than 131,000 animals including about 58,000 cattle and 22,000 deer were destroyed. These epizootics originated with importation of infected animals (cattle and pigs), contaminated hay and straw, unsanitary hides and wool. Nearby outbreaks occurred most recently in Mexico in 1946 (source: imported Brazilian bulls) and in Canada in 1952 (source: possibly clothing of foreign farm workers). Only strict regulatory measures keep us free of this disease. A disease-free buffer zone is maintained in Central America. Also free are: New Zealand, Australia (since 1872), Japan, and a few other countries.

CLINICAL SIGNS

After an incubation period of two to five days (one to 18 also reported) fever, profuse salivation, and lameness occur acutely. These signs mimic several other vesicular virus-induced diseases of livestock. Surviving animals become debilitated and suffer secondary complications including exungulation (sloughing of the hooves), gastroenteritis, abortion, mastitis, muscular degeneration (especially cardiac degeneration), and septicemia. Morbidity may reach 100% with natural mortality running usually about 5%, although occasionally as high as 50% (with muscle-myocardium attacking strains). Fallow and red deer are considered poor disseminators of the disease. White-tailed deer were involved in the United States epizootics.

PREVENTION AND CONTROL

In countries where FMD is not enzootic, government regulations are strict. In the United States, regulatory procedures include rapid laboratory diagnosis and subtyping, quarantine of infected premises, slaughter, and sanitary disposal of all infected and susceptible in-contact animals, inspection of adjacent premises, disinfection of affected buildings and real estate. Restocking of property with "sentinel" animals after a suitable barren period, usually about six weeks, tests the efficacy of the eradication program.

Although FMD is exotic to this country, knowledge of its signs and destructive potential will give the importer some patience with bureaucratic red tape.

TREATMENT

Not applicable. (2. Cottral & Callis, 157)

MALIGNANT CATARRHAL FEVER (MCF) is a sporadic viral disease of deer, cattle, and buffalo. The virus has been isolated and proven to be a herpes virus on the African continent, but never identified elsewhere. The exact method of transmission is not clearly understood. Blood-sucking insects, sheep, and the African black wildebeest may be intermediate silent carriers of the virus. Sheep and wildebeest may show vague signs but are generally asymptomatic shedders of the virus, while the sensitive species have a high death rate. In New Zealand, a direct deer to deer, but not deer to domestic livestock, transmission is suspected. In New Zealand, MCF may cause about 50% of all farmed deer losses. In Australia (New South Wales), fallow deer exposed to infected rusa (*Cervus timorensis*) and chital (*Axis axis*) deer have not become infected. (7. English, 95) In the laboratory, the rabbit has a high rate of infectivity with similar clinical sequellae. While natural field cases are unknown in the rabbit, the rabbit has become the experimental model. Also, rabbit-to-rabbit transmission, although attempted in the laboratory under intimate housing conditions, has never occurred. The rabbit is an experimental model only, and has not been implicated in natural spread among deer. (6. Aug/Sept '86, Massey, 49)

CLINICAL SIGNS

Five forms of MCF exist:

1. the peracute form leading to hemorrhagic gastroenteritis,
2. the intestinal form characterized by diarrhea and congested nasal and oral membranes,
3. the head-and-eye form including serous, then mucopurulent nasal and ocular discharge, crusting of the mouth and muzzle, respiratory distress, and blindness,
4. the chronic form consisting of cachexia (chronic progressive emaciation) and head-and-eye signs in a somewhat active creature,
5. the mild form seen in sheep, wildebeest, and certain resistant individuals of an otherwise susceptible population.

The incubation period is two weeks to one month. Occurrence is greatest during winter and early spring. The usual sequence in deer is anorexia (loss of appetite) and pyrexia (fever) followed by diarrhea, later, with hemorrhage. Some deer may appear blind and show some other head signs. Death occurs within two or three days of the onset of signs. Symptomatic treatment with fluids and antibiotics may prolong the agony a few more days.

PREVENTION AND CONTROL

MCF is a disease of farmed deer; it is unrecognized in feral deer. Decreasing stock density will help reduce disease potential.

Better nutrition in winter and early spring, the peak seasons for MCF, may increase animal resistance.

Isolation from possible intermediate asymptomatic carriers (sheep!) is prudent.

No vaccine is available.

TREATMENT

Symptomatic, but generally unavailing.

Deer **POX**, a type of "scabby mouth," is most probably a deer-specific viral disease. The virus (a parapoxvirus—like ovine orf and bovine pseudo-cowpox) had appeared only sporadically until 1985. Since then, reports of outbreaks in New Zealand farmed deer have increased, while probably a low-level of disease has existed in the feral population all along.

Transmission is similar to that of the other parapoxviruses. Skin abrasions and cuts are susceptible to airborne attack of the virus via dust particles. Biting insects or direct contact with an affected animal will also spread the virus.

CLINICAL SIGNS

Scabby lesions on the mouth, head, and velvet form, then rupture and heal, in the more common form of the disease.

In the less common form of the disease, lesions start at the feet and work their way up only as far as the brisket. The death rate is high in this form of the disease.

PREVENTION AND CONTROL

The pattern of clinical disease followed by immunity is likely to mimic orf in sheep. Likewise, a crude vaccine, made from a homogenate of deer-pox scabs applied to a scarified skin area and repeated annually, would probably be as effective as the similar vaccine for sheep.

TREATMENT

Symptomatic.

Possibly, vaccination in the face of an outbreak, as in ovine orf. (6. Feb/Mar '86, Massey, 25–7; Aug/Sept '86, Massey, 46–7)

RABIES is a fatal viral disease of mammals. This highly contagious disease, caused by a relatively easily inactivated rhabdovirus, generally spares lagomorphs (rabbits) and rodents. Spread is most often through saliva contact of abraided skin surfaces, usually via a bite wound. Aerosol dissemination occurs only rarely (for example, inhalation of contaminated air in a cave full of rabid bats). Incubation period is variable, usually two to eight weeks although sometimes shorter, but occasionally much longer, to one year.

CLINICAL SIGNS

The rabies virus attacks the central nervous system and causes an encephalomyelitis, a combination of head signs and other nervous system signs. A furious and a dumb form of the disease exist sequentially in some species (dogs: dumb first, furious later) or singly in other species (ruminants: dumb form predominates). Death follows usually within ten days of the onset of signs.

Rabies in deer is rare, but has been reported sporadically. In Alaska, rabies is endemic in the wildlife (fox) population. The following describes the clinical signs exhibited by Alaskan reindeer in a rabies epizootic.

"In 1982, reindeer herders on the Seward Peninsula of Alaska reported seeing reindeer with signs of illness including incoordination, weakness, posterior paralysis, head weaving, confusion, and aggression. Affected reindeer were observed having trouble keeping up with the rest of the herd. These animals had blood around the dorsal portion of their head near the antler buds, blood streaks along their back, congested eyes, and tufts of hair missing. They appeared confused and uncoordinated, and when examined closely, became aggresive. A lactating reindeer attacked a herder during a roundup in June 1982; this animal appeared disorientated, slightly uncoordinated, had multiple lacerations and bruises around the head, blood streaks along its neck and back, and missing tufts of hair. It used its head and front feet to batter at the herder but returned to the herd after being forcibly repelled. While with the herd it continually attacked other reindeer with its head or front feet; no biting attempts were observed." (7. Dieterich, 99)

PREVENTION AND CONTROL

Vaccination of cats and dogs routinely, and livestock as needed.

Control of the wildlife reservoirs—bat, bobcat, fox, coyote, raccoon, skunk, and other wild carnivores. Chipmunks, gerbils, guinea pigs, hamsters, squirrels, mice, rats, other rodents, and lagomorphs (rabbits and hares) are generally considered to be relatively resistant to rabies.

TREATMENT

None.

OTHER VIRUSES Deer can be mildly affected by several other viruses that are usually associated with other species. Bovine virus diarrhea (BVD) causes serological conversion in deer and at times mild gastroenteric disease (3. 1981, Fletcher), mild upper respiratory signs, and foot problems (7. Dieterich, 100). Evidence of other bovine respiratory viruses (infectious bovine rhinotracheitis/IBR, Parainfluenza 3/PI3) appears on routine serological testing, although clinical signs are mild to inapparent. Another respiratory organism, the virus-like rickettsia that causes Q-fever can be carried by deer. The tick-borne flavivirus of louping ill encephalitis has been found serologically and clinically in Scottish deer. (3. 1981, Fletcher) Herpes viral antibodies and keratoconjunctivitis have been reported from several sources due to a deer-adapted herpes strain. (3. 1985, Bonniwell) *Equine Herpes Virus I* has been reported from serology and isolation procedures involving fallow deer and wild ruminants. No clinical disease was detected. Abortion and neurological signs, however, can be associated with this virus: zebra abortions, alpaca blindness, antelope neurologic signs. (Dubovi, pers. comm.) Rotavirus coinfection occurs with colibacillosis.

General References: 1, 5, 6, 7, 8 10, 11, 15.

MISCELLANEOUS PROBLEMS

ACCIDENTS and **TRAUMA** are usually a result of man's faulty manipulation in farmed deer. Some pointers on the art and science of shepherding fallow deer (which must be handled quite differently from red deer) are contained in other chapters of this book. Other advice on handling facility construction (crushes, gates, holding areas, and raceways) should be studied and evaluated with your own layout in mind. Common sense indicates calm handling of this excitable species. Meticulous carpentry is a must; no exposed nails or wires, no gaps in fencing or at gates.

BLOAT is a potential complication in ruminant anesthesia. See *ANESTHESIA* **Positioning**. Eructation of ruminal fermentation gases is an intermittent and necessary life function. If ruminal gases are not eliminated, pressure exerted first on abdominal organs will progress to thoracic compression, respiratory embarassment, and finally death if left uncorrected. Procedures requiring anesthesia should be performed on fasted animals if possible. The anesthetized ruminant should also be positioned to minimize esophageal occlusion.

Bloat due to ingestion of rich feeds or a sudden change of diet also occurs in ruminants, although it does not appear to be a significant problem in fallow deer. Treatment generally involves oral medication, with trocharization a last resort. Consult your veterinarian.

PINKEYE (keratoconjunctivitis) in deer follows a clinical course similar to the disease in cattle. To date, however, an infectious agent has not been isolated from deer. The disease has been reported from New Zealand and Alaska (reindeer). Dusty conditions and insect strike are considered to be contributing factors. Control and treatment should follow guidelines already established for cattle. (7. Dieterich, 99; 15. van Reenen, 134)

PNEUMONIA in farmed deer is rare, but occurs when environmental, infectious, and stress factors coincide. Opportunistic bacteria already present in the lung (example, *Pasturella multocida*) can then do their worst. Any shelter/ventilation problems should be solved, stresses (crowding, nutritional, etcetera) reduced, and infectious organisms treated, if possible.

REGULATORY CONSIDERATIONS—When entering the breeding stock market, be prepared for the expenses and time-lags involved with the shipment of domestic animals, especially a relatively unknown species. To obtain the smoothest results, involve your veterinarian early in your plans.

As a general rule of thumb, whenever animals move, they always require some kind of paperwork and very often must be blood and/or TB tested.

Regulations are generally least stringent for **intrastate** movement.

The next level of difficulty is **interstate** movement. Some states do not have well-defined rules for fallow deer. This can engender an additional delay while the state's authorities confer and decide policy.

Most stressful on the animal, you, and your veterinarian is **foreign** export. Fallow deer breeding stock will generally require more extensive testing than cattle of the same class. Each country will vary, of course. Months of advanced notice can be necessary to complete required tests. ***Estimate!*** delivery dates conservatively, and only do so after consulting with your veterinarian.

The flip side of the coin is your purchasing and importing of deer. Again be prepared for bureaucracy and red tape. Make your arrangements well in advance to minimize disappointing delays—like, possibly the loss of a breeding season.

Remember, however, there are reasons for stringent testing regulations. The rules are for your (the deer farmer's) economic protection, as well as for the protection of the United States domesticated animal population. The larger your herd, the more you have to lose if disease is introduced. So, if someone offers you a possibly illegal shortcut, ask yourself "Why?" and refuse!

SELENIUM is a trace mineral which may exist in a deficit, normal, or excess level in the local soils. Its presence in forage is related to soil selenium content.

Selenium in excess is exquisitely toxic. In deficiency, a spectrum of problems can occur including white muscle disease, reproductive malperformance, immunosuppresion.

Supplementation via oral or parenteral route is indicated if the deer are deficient. Supplementation is strictly regulated by law because of the very toxic nature of this mineral if in excess. Regular selenium blood testing may help the veterinarian to round out a balanced nutritional program.

Data: Blood Selenium (ug/dl)
NORMAL: horse—15–35
cow—10–30

LUCKY STAR RANCH—before supplementation—does/fawns 1/86

20.3	4.2
10.5	2.9
9.7	2.9
7.2	4.5
8.9	6.1
5.6	

LUCKY STAR RANCH—after supplementation—bucks 6/86

19.7	21.5
20.7	25.5
20.6	20.8
19.3	18.5
26.0	25.1
23.7	25.5

There are no established normals for fallow deer at the diagnostic laboratory that we use.

STRESS-GASTROENTERIC-HEMORRHAGIC-SHOCK SYNDROME The cervine gut is the target organ for many infectious and noninfectious diseases, which all have as their primary clinical sign bloody diarrhea. Often other clotting problems are also associated.

"Haemorrhagic diseases of deer are a group of infectious or non-infectious diseases characterised by haemorrhagic enteritis, serosal haemorrhages, myocardial haemorrhages, ulcerations of the oral cavity, ocular involvement, and/or coronitis. Whatever the initiating factors, there results a combination of cascades of clot-forming and clot-lysing reactions. In those diseases of viral aetiology, endothelial cell damage leads to disseminated intravascular coagulopathy (**DIC**) with eventual consumption of clotting factors, haemorrhage, defective haemostasis, and shock. Endotoxins from gram-negative bacteria may similarly lead to DIC. Additional non-specific factors such as stress, transport, postcapture myopathy, and heavy metal poisoning may result in typical signs of haemorrhagic disease." (7. Morton/O'Hara, 476)

The infectious causes of cervine hemorrhagic disease complex include: bluetongue, EHD, BVD, MCF,* yersiniosis, enterotoxemia and others, like the exotic diseases FMD* and rinderpest.

**See* VIRUSES.

STRESS/SHOCK—POSTCAPTURE MYOPATHY When surprised, cornered, and threatened, animals experience a fight-or-flight response. This self-preservation alarm system involves the body's thymicolymphatic system. The stress of capture can lead to shock and/or myopathy (muscle necrosis). Onset of signs can vary from immediate to several weeks after the insult.

Four different sydromes exist:

1. Peracute death
   –within minutes of capture
      or
   –6 to 48 hours after capture (Upon disturbing normal-appearing animals, excitement and death occur within 1 to 2 minutes.)
2. Acute death
3. Ataxia myoglobinuria
4. Muscle rupture (14. Spraker, 708-9)

"Although some of the animals, both captured and myopathic, were alert in appearance after capture, many were very exhausted. A 'wry' neck often developed within 45–60 minutes after capture. Some animals showed either knuckling over of the fore or hind legs, dragging a leg, or reluctance to move within hours. Paresis developed within 24 hours and progressed to a general paralysis. Even when assisted to stand, lack of control of the legs was apparent. In acute cases the deer became recumbent after capture and died within a short time.

Deer often urinated soon after capture, with the urine being a very dark, almost black, coffee colour." (7. McAllum, 68)

Hypo or hyperthermia will exacerbate this syndrome.

It is important to realize that stresses other than chase and capture can trigger the shock sequence.

Treatment is generally unrewarding so prevention should be the goal. Treatment of acidosis and shock, however, may be helpful in less serious cases.

WINTER DEATH SYNDROME is a euphemism for the malnutrition/starvation complex. Deer are naturally lean with little fat store as a ready source of energy during times of cold and nutritional stress. (15. van Reenan, 120).

The reason there is a hunting season for our indigenous white-tailed deer is to minimize winter death syndrome in the wild population. The harshness of winter with its finite browsing and foraging potential can take an enormous toll.

Affected animals may die suddenly for no "apparent" reason or may waste and then die. Upon close inspection, dead animals are grossly thin but often appeared plump because of their thick, sometimes starry (a bad sign) winter coats. Also, these animals may have rounded bellies filled with bedding-quality feed. This lends a deceptive appearance in cases of malnutrition (versus starvation). These malnourished animals have had a *sufficient quantity* of feed to eat, but nutritional *quality was not adequate.* (In true starvation, quantity of feed is also a problem.) On post-mortem examination, body fat depots (in bone; and around heart, intestines, and kidney) are diminished or lacking.

The effects of other health problems (like disease or parasitism) are additive to nutritional stress.

Supplemental feeding to balance the ration makes this syndrome more controllable in closely-observed farmed deer herds. However, the nutritional balance of free-ranging herds (example, most reindeer herds) is still dominated by the vagaries of climate, disease, and feed availability.

References *HEALTH AND VETERINARY MANAGEMENT*

1. **Aglink Information Sheets.** Wellington, New Zealand: Ministry of Agriculture and Fisheries Information Services, 1981-1985.

2. Amstutz, HE, ed. **Bovine Medicine and Surgery 2nd ed.** Santa Barbara: Am Vet Pub Inc, 1980.

3. **British Veterinary Association Congress: Papers.** Exeter, 1985.

4. Chapman, DI and N Chapman. **Fallow Deer: their history, distribution and biology.** Terence Dalton Ltd, 1975.

5. **Deer Course for Veterinarians: Proceedings.** Deer Branch/NZ Veterinary Association, 1981-1986.

6. **The Deer Farmer** magazine. Wellington North, NZ: Deer Farmer Publications Ltd, 1983-1986.

7. Fennessy, PF and KR Drew, eds. **Biology of Deer Production [Proceedings: Intl Conf Dunedin, New Zealand 13-18 February 1983].** Wellington, NZ: Royal Society of NZ Bulletin No 22, 1985.

8. Fraser, CM, ed. **The Merck Veterinary Manual 6th ed.** Rahway: Merck & Co Inc, 1986.

9. Georgi, JR. **Parasitology for Veterinarians 4th ed.** Philadelphia: WB Saunders Co, 1985.

10. Kahrs, RF. **Viral Diseases of Cattle.** Ames: Iowa State Univ Press, 1981.

11. Merchant, IA and RD Barner. **Infectious Diseases of Domestic Animals 3rd ed.** Ames: Iowa State Univ Press, 1964.

12. Reinken, G. **Grun-und Brachlandnutzung durch Damtiere.** Angregungen fur Produktion and Absatz, 10, 1977.

13. Soma. LR, ed. **Veterinary Anesthesia.** Baltimore: Williams & Wilkins Co, 1971.

14. Spraker, TR. "Capture Techniques in Feral Ruminants." **Vet Clin North Am [Large Animal Pract]** 1986; Vol 2: No 3, 693-710.

15. Yerex, D, ed. **The Farming of Deer.** Wellington, NZ: Agricultural Promotion Associates Ltd, 1982.

20

Roundup

Fallow deer are more timid than domestic red deer. Their flighty nature and their tendency to panic easily have to be considered when herding the deer into roundup yards.

When it is time to round up the herd, you should develop a lure with feed, plus a reward reinforcement system, well ahead of time. When fallow deer are used to being handled and know a special person well, they are easily baited into an area. Feed them in the same area where you plan to work them, and choose a small pasture or yard to reduce the possibilities for getaway.

Once you start working them, keep them in small paddocks, since they tend to get nervous and flighty. If you have provided for a race between your pastures, bait them into that race and then walk them down to the yards with a piece of black silo plastic, a tarp or some sacking. This piece of plastic should be approximately six feet high and run across the whole race. Three or four people can direct the deer in almost any direction. Do not let them find a little hole to look through or, even worse, slip through. They will definitely take any chance for escape. If one goes, they will all go, and can't be stopped. So, get everything ready and in order before you have your first go, because your preparation will decide if it works the first time or if you have to herd them over and over. It is far less stressful for man and animal if you spend extra time preparing and planning and avoid repeat herdings.

Trying to push deer from a large pasture into a small yard with force is almost impossible. Even if you succeed with the first bunch, you might spend the rest of the week with the last five deer. Before you bring your deer into close confinements or yards, you should have taken your bucks out of the herd. If bucks have antlers, they will injure or perhaps kill some Does or fawns as soon as they get excited when they are crowded together.

There are several ways to get bucks out of your herd. Try well in advance to bait the bucks into an extra paddock, even if you have to take some of the Does with them. Does can later be released. Another possibility is to use a creep, which is quite easy to install. The spacing of 14 inches between the "2 X 4's" and the inside height of 35½ inches prevent most of the bucks from slipping through.

Only a few two-year-old bucks went through those creeps at Lucky Star Ranch, which is not so bad, since they are not as aggressive as the older bucks. If you have to, you can even harvest these two year olds and market them together with your yearlings. The meat quality is still quite good.

Any use of a tranquilizer gun to immobilize the bucks, if it becomes necessary, should be performed by people who know exactly what they are doing. Information on tranquilizers should be studied with great care. Drug label dosages and warnings should be observed with extra care in food-producing animals. Any meat/slaughter withdrawal times must be strictly observed. Work closely with your veterinarian.

More than 25 years' practice in immobilising wild animals has proved that there can be no uniform dart for the whole range of veterinary use. The large number of different syringe systems and sizes available on the market underlines this.

Engineers and field experts have now developed a universal CO^2 gasrifle with interchangeable barrels for all standard dart types and kinds of application. This gasrifle is especially economical since expenses result mainly from the wear of parts.

For wildlife parks, zoological gardens, animal-preserves and scientific research in the wild a widely variable precision projector of robust design is necessary in order to meet all requirements.

The CO^2 gasrifle G.U.T. 50 is a pressure projector for immobilisation, distance injections and marking of game, domestic animals and birds.

It can be provided with barrels of different calibre and length from which light weight darts can be fired gently.

Plug-in barrels allow the use of especially small dimensioned syringes for thin skinned animals or birds.

Rapid sliding pressure adjustment allows safe control of firing distances from 2 to 50 metres. It is soft triggered. A detachable standard telescopic sight or Aimpoint sight assists target finding.

At a special manometer the most favourable pressure/distance relationship can be read.

The rifle is easy to maintain and has been tested under all climatic and field conditions.

Back to the roundup. Remember, if possible never force the deer. They will turn and run against you and nothing will stop them. It is easier to wait one-half hour until they finally make up their minds to go through a gate than to have them panic. When reassembled, they will be even more excited.

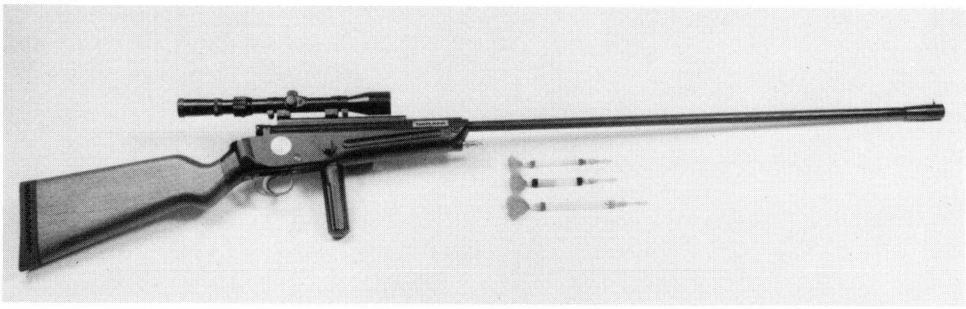

TECHNICAL DATA:

Calibre:	6, 8, 11 and 13 mm	Length:	1300 mm
Pressure range:	2-10 bar (resp. 2-25)	Width:	60 mm
Pressure camber volumes:	125 cm^3	Height:	175 mm
Effective range:	50 m	Gas supply:	16 gr CO^2 capsules for 10 shots of each 5 bar (for CO^2 or NO^2 capsules of different sizes and standards adapter rings are available).
Operational range:	70 m		
Weight:	2.5 kg		
Dart systems:	Telinject, Distinject, CAP-CHUR		

For manual operation an additional car type valve is provided.

REASONS FOR HAVING A ROUNDUP:

Weaning of Fawns
Stock Census
Fawn Census
Herd Liveweights (especially fawn)
Eartagging and Tattooing
Health Tests and Preventative Injections (See Health Chapter)
Blood Tests for Metabolism Profile (See Health Chapter)

WEANING OF FAWNS

When to wean is a topic that many world experts are debating. Our experience may illuminate that debate, but still, there are no pat answers to the questions. Some years ago, we weaned our fawns in September, but concluded that September (in a location near the Canadian border) was too early. At that time, we did not have good control over our breeding season and we had too many young fawns, some of whom were only a few weeks old at roundup time.

For that reason, we changed to a later weaning date. The late fawns should not have been taken away from their mothers in September, as some of them did not make it through the winter. We then did our roundups in December, but that is our busiest time selling venison, so the roundup did not happen much before the beginning of January, when it is quite cold to work outside here, and there is often too much deep snow. We lost quite a few fawns between October and December because of bad weather conditions.

From late October until the middle of December, during the rutting season, we try not to disturb the animals. The rut can be affected, and, with the adult bucks in the herd, problems may arise if they become agitated. I now tend to go back to an earlier schedule, and to do our roundup in the first two weeks of October, just before the rut. Since we remove our bucks from the herd by the end of December, all fawns are at least three months old or, in most cases, four or four-and-one-half-months old by roundup time.

STOCK CENSUS

Counting your stock is important for many reasons. You need a census of your breeding Does. You started with a certain number and you have to know if any are missing. There could be different reasons for missing stock. One possibility is a hole in the fence. A good deer farmer will check his fences regularly and make sure that animals cannot escape. Animals could have been taken by poachers. There could have been natural deaths among the older Does, but normally one would have found them during the winter or early spring. That is the time when the old Does usually die. There could be disease (even though fallow deer are fairly disease resistant), and you have to be aware of that danger at all times. A serious loss problem could occur with deficiency in trace minerals, like selenium, for example. At the Lucky Star Ranch, we lost animals due to this deficiency. See the chapters on feeding and health. Whatever the reason for missing animals might be, stock has to be counted, and the right time to do that is the yearly roundup.

FAWNING CENSUS

This is the critical area for profit. The fawn crop will decide the profit or the loss. When you count your fawns and find that you had an 80 or 90 percent fawning rate, you know that you are doing fine and that you are already halfway down the road to profit. The roundup is the "moment of truth" for every deer farmer, because it is the first time when he knows for sure how the year is going. When you are below a 70 percent fawn crop, you have a problem, and it will be difficult to call it a money making year. Check into all possible causes and try to find the answer as quickly as you can. Annual returns are quite similar to beef cattle farming. There is only one crop per year and it had better be good.

HERD LIVEWEIGHTS
(especially the fawns):

Weighing Does is mainly for statistical purposes. It can, however, be a barometer of your feeding program. Adult Does should weigh between 90 and 110 pounds, yearling Does between 70 and 90 pounds, and fawns between 40 and 55 pounds at the age of approximately four months. If you can say that you had a good year with a good fawn crop, all you have to worry about is how you get them through the winter in excellent condition, so that they will start to grow as soon as you turn them out to pasture.

EARTAGGING AND TATTOOING

We have found it practical to eartag our deer. We started with round little sheep tags with numbers, and for every year, we have a different color so that we can see from a distance how old an animal is. Some people will have you believe that they can tell age simply by looking at an animal, but don't buy that for a minute. Fallow deer are difficult to age and a colored tag cannot lie. So far, we have kept the color code but we are changing to the square sheep tag with bigger numbers so that we can see the numbers better from a distance. Because deer tend to lose their tags quite easily, we have also adopted tattooing so that we can accurately re-tag the animal if it has lost its tag. We tattoo the same number as the eartag, with a year letter, the same as the beef industry. 1986 = U, 1987 = W, 1988 = X, 1989 = Y, and 1990 = Z. In 1991, it will start with A again, and so on. The letters I, O, Q, and V should not be used, because they can easy be misread. We use cattle tattoos, the same ones we are using for our Polled Herefords.

HEALTH TESTS AND PREVENTATIVES:

LUCKY STAR RANCH FALL ROUNDUP

Dewormer	Vaccinations	Vitamins (if needed)
	Clostridium-8	Vitamin E/Selenium
	Leptospirosis-5	

THE REST OF THE YEAR AT LUCKY STAR RANCH

Coccidiostat (if needed)	Fecal examination periodically
Dewormer periodically	Nutritional balance = goal

BLOOD TESTS:

AT LUCKY STAR ROUNDUPS

We take a small sampling from representative classes of animals (buck, Doe, fawn).

Selenium blood levels

—to monitor herd's supplementation in our selenium-deficient part of the country

Serum disease titers

—primarily for research purposes to establish a baseline of fallow deer serum titers for various bacterial and viral diseases

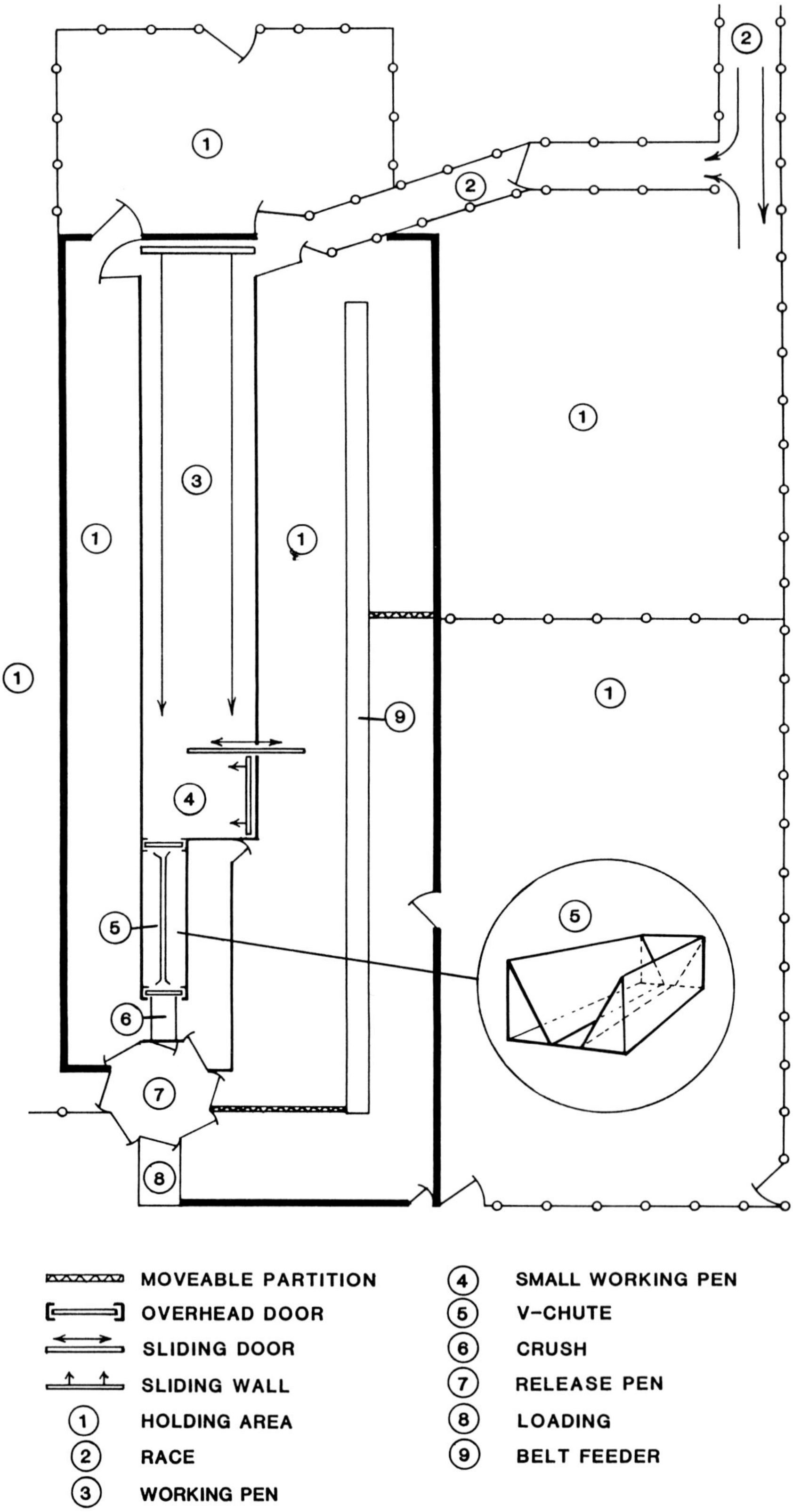
1
2
3
4
5
6
7
8
9
MOVEABLE PARTITION
OVERHEAD DOOR
SLIDING DOOR
SLIDING WALL
1 HOLDING AREA
2 RACE
3 WORKING PEN
4 SMALL WORKING PEN
5 V-CHUTE
6 CRUSH
7 RELEASE PEN
8 LOADING
9 BELT FEEDER

The deer are being walked down the race (2) with the help of a tarp, some silo plastic or sacking. It is important to keep the tarp straight and tight to the ground so the deer will not find a hole to escape. If they know they cannot escape, they remain quite calm.

Photo: Author

The deer enter the barn, run past the sliding wall (JR-wall) into the large working pen (3) The sliding wall can be brought all the way towards the sliding door, it runs on tracks.

Photo: Author

With the help of the sliding wall, an invention of the German deer farmer Josef Rüssmann, the deer can easy be brought down to the working pen and the sliding door.

Photo: Author

The sliding door helps to cut out a smaller group of deer which are kept in the working pen (4) from where they run singly into the V-shaped chute. The reason for the V-shape is, that deer don't seem to pile up in one corner as they do in narrow pens with straight walls, where they can easy be trampled on, especially if you have fawns and adult animals together in the roundup.

Photo: Author

At the end of the V-shaped chute is the crush, which can easily be adjusted to the size of each animal. Once the deer is comfortable, we release the trap door and the animal hangs in the V, without touching the floor and is totally immobilised. It cannot hurt itself nor the men working and is at no time under stress.

Photo: Author

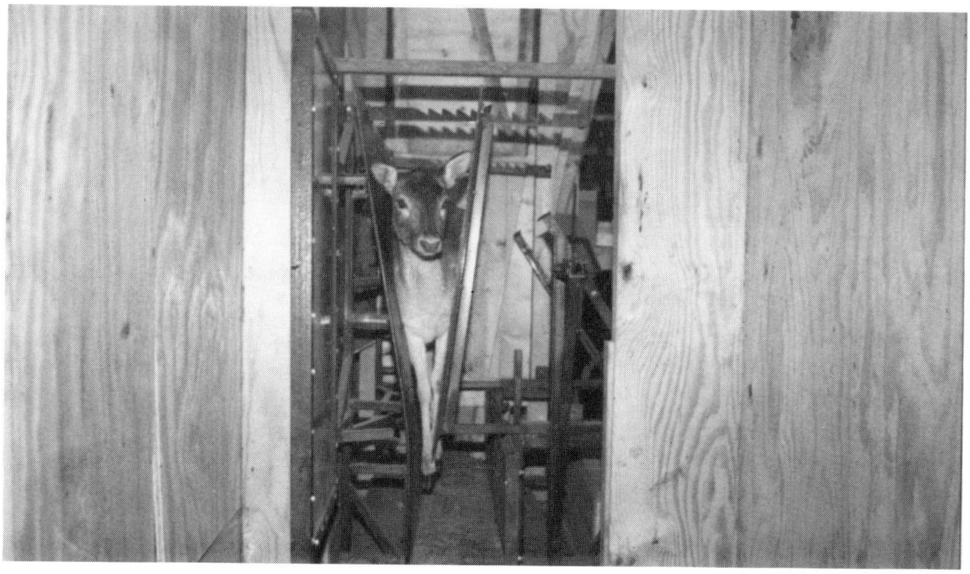

The idea of the two floors in the crush, the upper floor being a trap door, comes from New Zealand. The rest of the crush was designed by our mechanic Don Kilborn who is constantly improving it. This crush helps us to work between 200 and 250 deer per day.

Photo: Author

We have tried many different agricultural scale systems and have come to the conclusion that the TRU-TEST electronic system AG 300 series, is by far the best and the most reliable.

The TRU-TEST system can be used with or without a printer. Weights can be recorded manually as animals are weighed. A printer can be used to print out weights and identification information once all animals have been weighed.

A LCD Display
B Waterproof keyboard
C Serial interface Port
D Rugged Aluminium Case

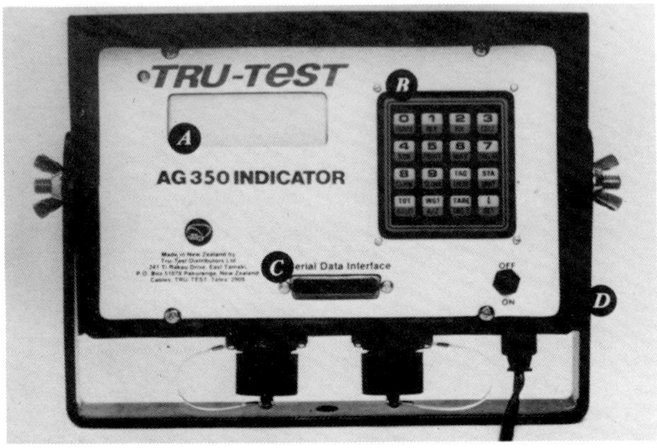

- 12v DC Electronics.
- Portable, showerproof, dustproof case.
- Large, easy to read liquid crystal display.
- Accurate to within ± 1%.
- Stable weight reading within 3 seconds even with moving animals.
- Unique non volatile memory, stores up to 1600 tag numbers and weights; information is retained even when power is disconnected.
- Computes and stores minimum, average, maximum and total weights, as well as counts the number weighed.
- Easy keyboard setup of scale features.
- Automatic taring of platforms, crushes or crates.
- Automatic zero tracking eliminates manual taring after each animal.
- Standard RS 232C Serial interface.
- Variable resolution settings — 50 grms to 5.0 kgs.
- Selectable baud rates 150, 600, 4800.
- Protected against polarity reversal.
- Keyboard functions can be controlled via serial interface allowing remote control from computer terminal.

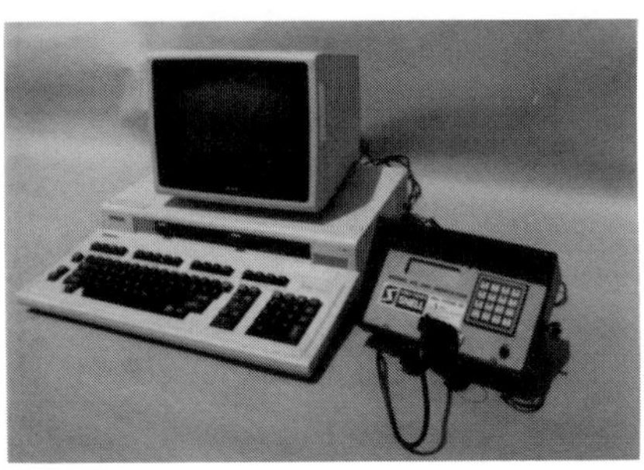

TRU-TEST offers a Software package that allows it to upload information directly from the indicator unit to an IBM compatible computer. TRU-TEST memory retains information even when power is disconnected.

21

Deer Farmer's Calendar

JANUARY FEBRUARY MARCH	Provide adequate shelter for all animals Feed corn silage or haylage and grain Feed hay ad libitum (free choice) Provide clean water at all times (heated waterers may be necessary) Take manure samples Worm deer in March twice within 3 to 4 weeks. Start pasture control measures in March, check all fences.
APRIL	Worm yearling bucks to guarantee maximum weight gain Continue fence control Release deer slowly into green summer pasture Be careful to allow a very small pasture at first so that the digestive tract can adjust to the change in diet More grass and less legumes is safer to begin with Bucks begin shedding antlers (End of April)
MAY	Make sure fences are fawnproof Fence off cover areas in middle of pastures Buck roundup after all antlers have been dropped
JUNE	Beginning of fawning season; do not disturb more than necessary Patrol fences for signs of coyotes and other predators Bucks begin growing new antlers Yearling harvest bucks need good pasture and supplements Provide good water, any necessary minerals, including, if necessary, selenium Harvest hay as early as possible, for highest nutritional content Observe fawning without disturbing

JULY	Fawning continues Worm yearling bucks (intended for harvest) again Bucks are still in velvet Good pasture is necessary to guarantee growth of fawns Bucks begin to gain weight, preparing for rut
AUGUST SEPTEMBER	Start creep-feeding fawns in pasture with grain (August 1st) Bucks have developed antlers and will rub off velvet Fawns are growing fast, and need lots of good forage Bucks are getting heavy and fat, necks swelling Bucks may begin to get aggressive Begin your marketing strategy for October
OCTOBER	(busiest month of the year) Harvest first group of yearling bucks Prepare and begin roundup (first two weeks of October) Count and wean fawns; work them later; keep in dry paddock with shelter Process Does, check for udders; make sure condition is good to ensure early conception Vaccinate Does: leptospirosis, clostridia, also deworm Take blood samples of all groups of deer, check for selenium Rut season begins; Does will stimulate bucks Does go into breeding pastures after roundup Bucks into breeding pastures (mating groups or single sire) Yearling Does in extra breeding pasture with older buck Watch out for aggressive bucks; Do not allow children into breeding pasture Never go into breeding pastures alone on foot; best to use a vehicle Do not disturb mating more than necessary Continue harvesting yearling bucks
NOVEMBER	Process the weaned fawns Count, weigh, vaccinate, eartag, tattoo, and separate by sex, select male fawns for polling Feed hay to fawns *ad libitum* (good 2nd cutting alfalfa) and water Start fawns slowly on grain again and raise slowly to one pound per day Harvest of yearling bucks continues Segregate late fawns and runts

DECEMBER	Begin to separate bucks and Does Bucks go into their own pasture Feed bucks well to regain strength after rut Bucks should be out by December 15 (December 31 at the latest) Breeding Does go into winter pasture Continue priority feeding of fawns until pasture time

NOTE: These dates will change considerably according to climate and location. The calender above is based on the climate at Lucky Star Ranch, located in upstate New York. Because of the long winter, we have to feed our animals six months of the year.

22

Recipes: the bottom line

Venison is often thought of as being tough and gamey, because it has often been handled wrong by hunters and households. There are two key points to be remembered when cooking venison. The first is fairly well-known, always serve any grilled or cooked venison medium rare and not "well-done." It can easily dry out and get tough once past the pink stage. The second key point, contrary to popular opinion, venison—especially of farmed young deer—is often best unmarinated. Marination is best used on tough parts like those of older animals. The strong flavor of most marinades tends to mask the delicate flavor of our farmed venison.

One of the simplest and best marinades is "hungarian style"—which is pure plain buttermilk. Plain yoghurt can also be used with equal success.

Here are some recipes which might help you enjoy venison a little more

FALLOW DEER SADDLE

"ILLIANA"

Wash whole saddle (bone in) under cold water, dry with a paper towel, take one or two skins off, take big roasting pan and put saddle in, sprinkle saddle with salt & pepper, put 2 quartered onions, 2 cut up tomatoes, 2 laurel leaves, some juniper berries all around saddle, melt some unsalted butter in another pan and when bubbling hot (not brown) pour over saddle. Put saddle in preheated oven (350 F), baste every ten minutes with juice in pan, (eventually add some red wine). After 50 to 60 minutes take saddle out of pan and wrap tightly in aluminum foil. Let rest for 10 to 15 minutes. Add more wine to all remaining things in pan and let it come to a boil. Scrape the bottom of the pan carefully, strain into smaller pot and let it boil down for a few minutes, then put in heavy cream, taste and adjust seasoning. (Illiana adds some cranberry jelly, usually 1 cube of beef broth and some ketchup, mustard & sour cream) Don't let it boil after mustard and sour cream are added. Carve saddle, pour juices from meat into gravy. Cut loins into smaller portions and put them back on the bone.
Serve saddle and gravy separately.

VENISON STEW

"HUBERTUS"

Cut 2 lbs. of meat from neck, brisket, front shoulders, leftovers from hindquarters or wherever, in diced pieces. You can buy it already diced from Lucky Star Ranch. Put in a deep dish, add salt, 1 cup of red wine, 2 cups of water and a little white vinegar: the liquid should completely cover the meat, if not add more wine. Cover bowl with aluminum foil and allow to marinate in a cool place for 2 days, from time to time turning the meat. Remove the meat from the marinade, dry with paper towels, and roll in flour. Dice 3 slices of bacon and place in a casserole, braise until slightly crisp; add a little butter, then quickly braise meat chunks. Spice with pepper, 2 cloves, 2 laurel leaves, 1 finely chopped onion and 2 finely chopped small dill pickles; stir and add about ½ cup of the marinade. After stewing for around 60 to 90 minutes, thicken sauce with a mixture of cold red wine and flour. Eventually add salt if needed. The sauce should be spicey. In Germany this meal is often served with fried potato pancakes and a green salad. Also with "spaetzle" and apple sauce.

Upon completion immediately drink all remaining wine in bottle!

VENISON ROAST

"LUCKY STAR"

Clean and trim hindquarters. Rub in salt and pepper and sprinkle with a bit of thyme or sage. Place 2 or 3 slices of bacon on roast and place in a pan with several onion quarters. Put hindquarter in preheated oven (375 F) and allow to cook for 60 to 90 minutes. During this time the roast should be frequently basted. From time to time red wine and/or water should be added. After removing from oven wrap in aluminum foil and let it rest for approximately 10 minutes.

Gravy: After removing roast from pan add about 1 pint of heavy cream and stir with drippings from roast. Scrape pan well so all residue from roast is dissolved in sauce. Season with salt and pepper after tasting. According to taste red wine and thyme or sage can be added. Sometimes it is helpful to use 1 or 2 drops of "gravy master." Place this mixture in a saucepan and cook (not simmer) for about 10 minutes until sauce is reduced to the proper consistency. Taste again and season accordingly. Serve gravy separately!

Leftover gravy and roast can be frozen. Leftover gravy can also be used as a starter for future gravies.

DR. MARY'S RECIPES

FALLOW VENISON MEATBALLS AND SAUCE

VENISON MEATBALLS

(small batch)

1 pound ground venison (*no* added fat)
1 egg
½ cup Italian-style bread crumbs (Progresso)
¼ cup grated Romano cheese (jar or better, freshly grated)
1 Tablespoon parsley
¼ cup milk

(large batch)

4 pounds ground venison (*no* added fat)
4 eggs
2 cups Italian style bread crumbs (Progresso)
1 cup grated Romano cheese (jar or better, freshly grated)
¼ cup parsley
1 cup milk

Mix well—1st combine egg and meat; then, stir in thoroughly bread crumbs, followed by cheese and parsley; then, add milk and knead to a uniform consistency. Small batch makes 6–7 meatballs and large batch makes 2–2½ dozen meatballs of approximately 2-inch diameter.

SAUCE FOR MEATBALLS

(small batch)

1-pound 12-ounce can tomato puree
1 bay leaves (laurel)
1 large clove of garlic (crushed)
½ small onion (chopped)
3 Tablespoons parsley (fresh) OR
1½ Tablespoons if dehydrated parsley
¼ teaspoon dill weed
⅛ teaspoon sweet basil leaves
⅛ teaspoon whole rosemary leaves
1½ Tablespoons olive oil

(large batch)

6-pound 10-ounce can tomato puree
6 bay leaves (laurel)
6 medium cloves of garlic (crushed)
1 medium onion (chopped)
½ cup parsley (fresh) OR
¼ cup if dehydrated parsley
1 teaspoon dill weed
½ teaspoon sweet basil leaves
½ teaspoon whole rosemary leaves
¼ cup olive oil

Stir ingredients into sauce.

Submerge raw venison meatballs into sauce (large batch) in an 8-quart pot with cover. Cook on medium heat for 10–15 minutes, then lower heat between medium and low for 1¾ hours. *Do Not Scorch Sauce!* Check and stir slightly every 15 minutes or so.

FALLOW VENISON MEATLOAF

1 pound ground venison (*no* added fat)
1 egg
½ cup plain bread crumbs
¼ cup grated Romano cheese (jar or better, freshly grated)
1 Tablespoon parsley
¼ cup milk
bacon strips or butter pats to top meat before baking

Mix well—1st combine egg and meat; then, stir in thoroughly bread crumbs, followed by cheese and parsley; then, add milk and knead to a uniform consistency. Bake at 350 in a glass baking dish to desired doneness.

FALLOW VENISON RIBS

350 OVEN TEMPERATURE
Approximate roasting time 30 minutes.

Wash the ribs in cold water. Place the ribs arched on a rack in an open roaster. Rub crushed **garlic** onto ribs. Shake on lightly **dill weed** and **crushed sweet basil leaves.** Add **fresh parsley.** Mill on **black pepper.** Cover with a layer of **sliced onions.** Top with strips of good quality **bacon** (to baste and protect this very low-fat venison).

Pour 1 cup of **dry red wine** (fairly good quality) over ribs when five minutes from being done.

Make sauce/gravy in the roaster from drippings, adding ½ pint of **sour cream** and more wine if desired.

FALLOW VENISON RIBS—alternate recipe

Wash.
Charcoal grill.
Turn ribs often, painting with barbecue sauce each time.

SADDLE OF FALLOW DEER VENISON

WHOLE SADDLE
350 OVEN TEMPERATURE
Approximate roasting time 15 minutes/pound, but check at 10 minutes/pound and be guided by meat thermometer.

Wash the meat in cold water. Place the saddle arched on a rack in an open roaster. Rub crushed **garlic** onto the meat. Shake on lightly **dill weed and crushed sweet basil leaves.** Add **fresh parsley.** Mill on **black pepper.** Cover with a layer of **sliced onions.** Top with strips of good quality **bacon** to baste and protect this very low-fat venison.

Roast to 120 on the meat thermometer, then pour 1 to 2 cups of **dry red wine** over the saddle (use fairly good quality wine). Finish roasting to **140 for very rare** or **150 for medium well.** Remove saddle to a cutting board and set for 10 minutes.

Make sauce/gravy in the roaster from drippings, adding 1 to 2 pints of **sour cream*** and 1 to 2 cups or more of the **wine*.**

Serve with egg noodles. Roasted chestnuts are a nice touch for garnish.

To carve, remove bacon/onions and set aside on platter, then carve and arrangemeat with bacon attractively.

**Vary amounts according to taste.*

FALLOW VENISON STEW

1 pound bacon (good quality)
5 pounds venison chunks (about 2-inch squares)
3 medium onions (chopped)
6–7 garlic cloves (crushed)
12–16 ounces sour cream
1–1½ cups dry red wine
½ cup fresh parsley (¼ cup if dehydrated)
1 teaspoon dill weed
½ teaspoon sweet basil leaves
¼ teaspoon whole rosemary leaves

Fry bacon. Put bacon aside. Place bacon fat in 6 to 8-quart pot. (Jenn-air french fryer/cooker is ideal—cook bacon and remove it leaving the fat.) Add garlic, onion, and other spices and cook for 2 minutes. Then combine venison chunks with spices. Grind pepper onto meat. Cook on medium to low heat for approximately ½ to 1 hour. Then add sour cream and wine—cook until sauce is uniform in consistency. Vary cooking time according to desired doneness of venison chunks—excellent rare (or well-done as in traditional stew). Add bacon for last 5 to 10 minutes to reheat. Serve with noodles, rice, or potato.

FALLOW VENISON STEW (JOAN'S)
(traditional beef-type stew)

5 pounds **venison chunks** (about 2-inch squares)
Brown meat in pan—in water, or fat of ¼ pound of **bacon.** Do not burn.
Pepper to taste.
3 **onions**—cook in (or 1 pound small whole frozen onions).
1 teaspoon **thyme**
Meat is done when brown.
Water (if needed).
1 pound sliced **carrots** or miniature carrots—chunked, washed, not peeled—cook till done.
3 large **potatoes**—chunked, washed, not peeled
Water to cover.
1–2 bullion cubes
Cook—more seasoning to taste. When the carrots and potatoes are done, stir in corn starch or flour with a little hot water, if needed to thicken.

NICOLA FLETCHER'S RECIPES

PRE-COOKED AND MINCED VENISON

VENISON BURGERS

1 lb minced venison (raw)
2 oz fine brown breadcrumbs
1 small onion, chopped finely
1 egg
Salt, pepper, juniper berries, ginger to taste

Mix all these together well and form into round flat cakes (they tend to bunch up when cooking so the flatter the better). They can be cooked in the oven or under the grill, but I think they are best seared in hot fat or charcoal grilled. Cooking time as for steaks.

VENISON PATTIES

1 lb minced cooked meat
1 tablespoon minced onion
Worcester sauce
1 tablespoon mustard
1 teaspoon sugar
White sauce

Mix the meat, onion and some seasoning together and form the result into little round cakes about 1 in. diameter. Brown them in butter and put them in a small casserole. Make a small amount of white sauce with dripping or butter, flour and milk, then add the Worcester sauce, mustard, sugar and some horseradish sauce if you like it more piquant. Pour this over the patties and bake for ½ hour in a moderate oven.

A SMOOTHER PATÉ

For every 1 lb of cooked meat, use 2 oz butter, and a small glass of port, sherry or good gravy. Put all these into a dish with a close-fitting lid and cook in a warm oven for 2 hours. Strain the meat and blend it thoroughly, adding as much of the juice as necessary to make a soft paste, remembering that it will stiffen a little when cool. Check the seasoning. Stir in a dessert-spoon of chopped parsley and one of chopped pickled pimentos, gherkins or a mixture of both. Put into small jars and cover with clarified butter if it is not to be used immediately.

Courtesy of Mrs. Nicola Fletcher, Reedie Hill Deer Farm, Auchtermuchty, Cupar, Fife, Scotland, KY 14 7HS; from her book Venison, The Monarch of The Table, *1983*

23

North American Deer Farmers Association

BY-LAWS

ARTICLE I

Name

The name of the organization shall be the North American Deer Farmers Association, Inc., incorporated under and pursuant to the Not-For-Profit Corporation Law of the State of New York.

ARTICLE II

Objectives

The objectives of the association are:

1) To establish and encourage friendly exchange of ideas on all matters concerning the breeding, handling and management of farmed deer among its members.

2) To promote deer farming in North America and to acquire and distribute to members information on all aspects of deer farming.

3) To provide a forum for discussion of deer farming problems and to keep members fully informed on all matters of interest to deer farmers and potential deer farmers.

4) To act as an industry association in presenting deer farming views to the Government and to Governmental departments.

5) To provide a registration system when the circumstances shall warrant.

6) To help ease regulations on inspections, permits and licenses; to extend the allowable areas for deer farming if restricted in certain states; to seek improvement of legal protection for deer farmers from illegal activities such as trespass and poaching.

ARTICLE III

Membership

Section 1
Any individual, firm, partnership, corporation or association involved in deer farming may become a member of this Association by making application therefor and by paying the dues as specified in Section 2 hereof.

Section 2
There shall be four (4) kinds of membership:

1) Dues of a Life Member shall be a one-time payment of U.S. $2,000.
2) Dues of an Active Member shall be $150.00 payable upon making application for membership and upon subsequent anniversaries thereof.
3) Sustaining Members are not obligated for the payment of dues.
4) Any individual, firm, partnership, corporation or association who has performed outstanding services on behalf of the Association can be recommended by the president of the Association as an Honorary Member and shall become such upon a majority vote of the voting members of the Association.

Section 3
Each Life Member and each Regular Member in good standing of this Association shall be entitled to one (1) vote on any matter and/or election brought before any regular or special meeting of this Association. A member in good standing must have paid all dues and other obligations owed by him or her to the Association.

Section 4
Any member having a grievance or complaint against the Association may appeal to the members of the Association at any regular or special meeting.

Section 5
Unless otherwise provided by law or these by-laws, all the rights, title, and interest of a member of this Association in or to the Association or its property cease on the termination of membership.

Section 6
All books and records shall be open for inspection by any member at the offices of the Association, located at the offices of the Executive Secretary.

Section 7
Any member of this Association who violates the By-Laws or rules and regulations of this Association, or who shall deceive or wrong the Association or a member thereof or other person, or who shall conduct himself or herself so as to make his Association undesirable or any member whose practices in the sale of deer are such as to impair the reliability or the records of the Association shall be censured, suspended or expelled by the Executive Committee after notice and a formal hearing.

The Executive Committee must notify a member in writing by certified mail of any impending action to censure, suspend, or expel said member. The member shall have 14 days from the postmark date of the notification letter in which to respond in writing by certified mail to the office of the Executive Secretary of the Association requesting a formal hearing before the Executive Committee. Such formal hearing to be held at a time and place designated by the President of the Association not more than 21 days after the postmark date of the member's reply letter. If the member fails to respond as specified above, membership shall be immediately cancelled. If a formal hearing is held, the Executive Committee shall take action and notify the member in writing within ten days of the action taken. There shall be no refund of membership dues paid.

ARTICLE IV

Section 1
Officers of this Association shall be: President, Vice-President, Secretary and Treasurer who are to be elected for terms of five (5) years by majority vote of all voting members present at the regular annual meeting of the Association. These officers shall perform the duties prescribed by the parliamentary authority adopted by this Association and those prescribed in these By-Laws.

Section 2
The Executive Committee shall by majority vote appoint an Executive Secretary at a salary established by the Executive Committee or until such time as he or she may resign.

Section 3
The President Shall:

a) Preside over all meetings of the Association and the Executive Committee.
b) Sign with the Treasurer all notes, deeds on other instruments on behalf of the Association.
c) Call special meetings of the Association and the Executive Committee.
d) Appoint the chairman and members of all other committees of the Association. (as defined hereinafter by Article VII, Section 1)
e) Serve as ex-officio member of all committees, except any nominating committee.

Section 4
The Vice-President shall:

a) In the absence, disability or refusal to serve of the President, perform all duties of the office.
b) Oversee and facilitate the work of all committees of the Association.

Section 5
The Secretary shall:

a) Keep a complete and accurate record of all meetings of the Association and the Executive Committee.

b) Perform such other duties as may be required by the Executive Committee.

Section 6
The Treasurer shall:

a) Sign all checks, and with the President, sign all notes and other obligations of the Association.

b) Receive and disburse all funds, and serve as the custodian of all securities and property of the Association.

c) Keep a full and accurate account of all financial transactions of the Association in books belonging to the Association and deliver such books to the successor in office when elected.

d) Prepare a full report of all financial matters and present such to the members at the annual meeting and to the Executive Committee whenever requested.

e) Deposit all monies of the Association in the name and to the credit of the Association in such depositories as may be designated by the Executive Committee.

f) Prepare an annual budget and present to the Executive Committee for approval and recommendation to the membership at the Association annual meeting.

g) Present to the Audit Committee all books, records, and other documents as may be necessary for a complete and accurate annual audit of financial books of the Association.

Section 7
The Executive Secretary shall:

a) Serve all notices required by law and these by-laws.

b) Retain custody of the seal of the Association and shall cause the same to be affixed to such instruments, on behalf of the Association, as shall be directed by the Council.

c) Keep a complete and accurate list of all members, their addresses and membership termination date.

d) Notify the officers and committee members of their appointment.

e) Perform such duties and responsibilities as determined by the Executive Committee.

Section 8
Candidates for the offices of President, Vice-President, Secretary and Treasurer shall be nominated by any member.

Section 9
Vacancies occurring after the annual meeting shall be filled in the following manner:

a) A vacancy in the office of President shall be filled by the Vice-President.

b) Vacancies in the offices of Vice-President, Secretary and/or Treasurer shall be filled by majority vote of all Executive Committee members at a properly called Executive Committee meeting, the notice for which must indicate that such an election is to take place.

Section 10
The Executive Committee may require certain officers to be bonded.

ARTICLE V

Meetings of the Association

Section 1
The regular annual meeting of this Association shall be held during the month of April each year, the specific time and place to be designated by majority vote of the Executive Committee. The purpose of the annual meeting shall be to receive reports from officers and committees to elect officers of the Association and to conduct other business as appropriate.

Section 2
Notice of the time and place for the annual meeting or any special meetings of the Association shall be given by the Executive Secretary by mail, at least 30 days prior to such meetings, to each member at his residence or place of business as it appears on the books of the Association.

Section 3
Special meetings may be called at any time by the President or Executive Committee. The President shall also call such meetings if requested in writing by ten or more members stating the specific business to be considered. The Executive Secretary shall mail the required notice to all members of the Association.

Section 4
A quorum of a properly called annual meeting or special meeting of the Association shall be those members present.

ARTICLE VI

Executive Committee

Section 1
The governing body for this Association shall be the Executive Committee. The Executive Committee shall determine the policy for the Association within guidelines established by the membership in attendance at the annual and/or special meetings of the Association.

Section 2
The Executive Committee shall be composed of the President, Vice-President, Secretary and Treasurer of the Association.

Section 3
Meetings of the Executive Committee shall be called by the President not less than three times annually, or more frequently if in his judgment the business of the Association requires it. Upon request by two (2) members of the Executive Committee, a meeting may be called at anytime, provided those requesting the meeting serve notice to the entire membership of the Executive Committee together with a detailed list of the objects to be considered or discussed. Notice of all Executive Committee meetings shall be postmarked by first class mail at least ten days in advance of such meetings.

Section 4
The Executive Committee shall control the business and affairs of the Association and make the necessary rules and regulations not inconsistent with the law or these by-laws for the management of the business and the guidance of the officers, employees and agents of the Association.

Section 5
A quorum for a properly called Executive Committee meeting shall be a majority of the Executive Committee membership.

ARTICLE VII

Committees

The Executive Committee or the President with majority approval by the Executive Committee shall have the authority to create committees as may be deemed necessary. All committees shall act under the authority and direction of the Executive Committee and shall report directly to the Executive Committee. The Vice-President shall be an ex-officio member of all committees except any Nominating Committee that may be appointed.

ARTICLE VIII

Parliamentary Authority

The rules contained in the current edition of "Roberts Rules of Order" shall govern this Association in all cases to which they are applicable and in which they are not inconsistent with these by-laws and any special rules of order the Association may adopt.

ARTICLE IX

Seal

The seal of the Association shall be in the form of a circle and shall bear the name of the Association and the year of its incorporation.

ARTICLE X

Amendments

These by-laws may be amended at any properly called regular annual meeting of the Association. Amendments must be approved by a two-thirds vote at a regular or special meeting of the Executive Committee preceeding the annual meeting. A two-thirds vote of all members present at the Association annual meeting is required for adoption of any amendments to these by-laws. Any member may present amendments for consideration by the Executive Committee.

PRESIDENT: Josef von Kerckerinck
VICE-PRESIDENT: David Vander Molen
TREASURER: Eliot C. Clarke
SECRETARY: Raymond Maguire

NADeFA address:
Lucky Star Ranch
Chaumont, N.Y. 13622

BIBLIOGRAPHY

Asher, G. W. Dr. 1986 Proceeedings of Vet. Comf. Deer Branch, NZ Vet. Assoc. Rotorua, July 1986, pp. 202–206

Asher, G. W. Dr. 1986 *Growth and Venison production: Red and Fallow Deer.* Ruakura Animal Research Station, Hamilton NZ

Adam, J. L. 1986 ditto.

Asher, G. W. Dr. 1985 *Oesrtous cycle and breeding season of farmed fallow deer.* Ruakura Animal Research Station,

Asher, G. W. Dr. 1986 *Inhibition of Antler Development of Fallow bucks. The Polling Technique.*

Asher, G. W. Dr. 1986 *Reproduction of farmed fallow deer (Dama dama* Ruakura Animal Research Station,

Amstutz, H. E. 1980 *Bovine Medicine and Surgery, 2nd ed.* Santa Barbara: Am Vet Pub Inc.

Acken, D. van 1972 *Zum Tagesrhytmus des Damwildes* Zeitung für Jagdwissenschaft 18, West-Germany

Aglink *Aglink Information Sheets,* Ministry of Agriculture and Fisheries Information Services, 1981–1985

Baker, K. 1973 *Reproductive Biology of Fallow Deer in the Blue Mountains of New Zealand.*

Bamberg, F. B. 1985 *Untersuchungen von gefangenschaftsbedingten Verhaltensänderungen beim Damwild,* West-Germany

Barner, R. D. 1964 *Infectious Diseases of Domestic Animals,* 3rd ed Ames: Iowa State University Press

Bogner, H., Prof. Dr. 1986 *Tierschutzaspekte hinsichtlich der Haltung von Damwild zur Grünlandnutzung,* Bayerische Landesanstalt für Tierzucht, Grub, West-Germany
U. Zeeb and K. Popp, dito.

British Veterinary Association Congress: Papers. Exeter, 1981 & 1985

Cabrera, Dr. Angel *Mamiferos Sud Americanos*

Chapman, Donald & Norma 1975 *Fallow Deer: their history, distribution and biology,* Terence Dalton Limited of Lavenham, Suffolk, England,

Doyle, William	1985	*Frontiers for Agriculture,* An Action Agenda for NY State, Arthur D. Little Inc., Henderson, Friedlich, Graf & Doyle, Inc.,
Drew, K. R.	1985	*Biology of Deer Production* Proceedings: Intl. Conf. Dunedin, NZ 13–18 Feb 1 Wellington, NZ: Royal Society of NZ, Bulletin No 22, 1985
Durden-Smith, Jo	1985	*Baron of Venison,* Connoisseur Magazine, New York, NY
Fennessy, P. F.	1985	*Biology of Deer Production* The Royal Society of New Zealand, Wellington, NZ
Fletcher, Nicola	1983	*Venison, The Monarch of The Table*
Fraser, C. M.	1986	*The Merck Veterinary Manual, 6th ed.* Rahway: Merck & Co. Inc.,
Georgi, J. R.	1985	*Parasitology for Veterinarians, 4th ed.* Philadelphia: WB Saunders Co,
Grzimek, Bernhard Dr. Dr. h.c.,	1972	*Grzimek's Animal Life Encyclopedia* Volume 13, Mammals IV, Kindler Verlag GmbH, Zürich, Switzerland
Hansen, Paul	1968	*Das Damwild,* Verlag Paul Parey, West-Germany
Hofmann, R. R. Prof. Dr.	1985	*Digestive Physiology of Deer,* University of Giessen, Giessen, West-Germany
Kahrs, R. F.	1981	*Viral Diseases of Cattle,* Ames: Iowa State University Press
McCall, Julie		*Effects of performance and prices on fallow deer profitability* Ministry of Agriculture and Fisheries, NZ
McCall, Julie	1985	*Fallow Deer Farmers Discussion Forum* M A F, NZ Te Awamutu,
M A F		*Information Service Deer Farming* Ministry of Agriculture and Fisheries, Wellington, New Zealand
Merchant, I. A.	1964	*Infectious Diseases of Domestic Animals,* 3rd ed. Ames: Iowa State University Press

NZDFA	1983–1986	*The Deer Farmer*, published by Deer Farmer Publications Ltd., New Zealand Deer Farmers' Association, Agpress Communications Ltd.,
Puttick, G.	1972	*Analysis of fallow deer population—on Groote Schuur Estate*, University of Cape Town, South Africa
Reinken, G. Prof. Dr.	1977	*Grün- und Brachlandnutzung durch Damtiere*, Landwirtschaftkammer Rheinland, Bonn, W-Germany
Reinken, G. Prof. Dr.	1980	*Damtierhaltung auf Grün- und Brachland*, with Prof. Dr. W. Hartfiel and Dr. E. Körner Verlag Eugen Ulmer, Stuttgart, W-Germany
Riney, T.	1955	*Evaluating conditions of free ranging Cervus elaphus*, N.Z.J. Science Techn.
Soma, L. R.	1971	*Veterinary Anesthesia* Baltimore: Williams & Wilkins Co.,
Spraker, T. R.	1986	*Capture Techniques in Feral Ruminants*, Vet Clin North Am (Large Animal Pract.) Vol 2: No 3, 693–710
Ueckermann, E. Dr.	1968	*Das Damwild*, Verlag Paul Parey, Hamburg and Berlin, W-Germany
Veterinary Association, NZ	1981–1986	*Deer Course for Veterinarians*, Proceedings: Deer Branch
Whitehead, G. Kenneth	1972	*Deer of the World*, The Viking Press, Inc. New York, NY. The Anchor Press Ltd., Tiptree, Essex
Weber, W.	1972	*Grzimek's Animal Life Encyclopedia* Volume 13, Mammals IV Kindler Verlag GmbH, Zürich
West Virginia University	1985	*Deer and Agriculture in West Virginia*, Cooperative Extension Service, Center for Extension and Continuing Education, Publication 812,
West Virginia Fence		Tightlock Deer Fence Brochure
Yepes, Dr. Jose		*Mamiferos Sud Americanos*
Yerex, David	1982	*The Farming of Deer*, Agricultural Promotion Associates Ltd., Wellington, New Zealand

INDEX

This book was designed by
DeCourcy Taylor
Terence Boylan was the Production Coordinator
The paper is Mountie Matte
The type is Caslon 540 and was
composed by DEKR Corporation,
Woburn, Massachusetts
The printing was by Daamen, Inc.,
Center Rutland, Vermont
The binding was by General Bookbinding,
Agawam, Massachusetts